A Beginner's Guide to Death and Disease

First published in Great Britain in 2016 by Chris Bradbury.

Revised 2018 and published by 13 September Publications

ISBN-13:
978-1539529521

ISBN-10:
1539529525

Cover Designed by Chris Bradbury

www.christopherbradburybradbury.co.uk

‘Anyone can stop a man’s life, but no one his death: a thousand doors open onto it.’

Seneca the Younger, 65A.D

death *n.* annihilation, bane, bereavement, cessation, curtains, decease, demise, departure, destruction, dissolution, downfall, dying, end, eradication, exit, expiration, extermination, extinction, fatality, finish, loss, obliteration, passing, release, ruin, undoing. *antonyms* birth, life.

disease *n.* affliction, aliment, blight, cancer, canker, complaint, condition, contamination, disorder, epidemic, ill-health, illness, indisposition, infection, infirmity, sickness, upset, virus

Chambers Dictionary of Synonyms and Antonyms (1989)

For Sally, George, Ellie and Frankie

Table of Contents

In the Beginning…

I have a theory.

It goes like this.

Everything we do is about death. Everything. From reading a book to eating to sexing (that's a real thing) to sitting at the back of the class in school learning maths. From the day we are born to the day we die, death is at the centre of all we do, either consciously or unconsciously. It is the basis of all fears, from spiders to heights to water. It drives us, gives us purpose and yet is equally as capable of rendering us inert, unwilling, demotivated and demanding of its presence in an effort to control it[1].

Sex is the transmission of life, the physical and psychological epitome of that selfish gene so espoused by Dawkins:

> 'We are survival machines - robot vehicles blindly programmed to preserve the selfish molecules known as genes[2].'

We like to think that it's a triumph of drunken fumbling on a Thursday night round town, but it's not. We boast about it, smile pensively next to the photocopier, dream of it. Yet, it's something far deeper, far more sinister than that. It is the monster that lurks within and, like an alarm clock, it wakes you up to the possibilities that come with gender difference. It is in our genes and there is nothing we can do about it. We are given an innate urge to procreate, to ensure the survival, not of the species, but of the individual. It is a craving that, if not met, can at the very least lead to a skewed vision of the world around us; at worst, it can lead us into temptations that rely upon the oppression of others to meet our needs. This gene-deep urge is so strong that, at least

---

[1] Quite possibly *Terror Management Theory*. Developed in 1986 by social psychologists Jeff Greenberg, Tom Pyszczynski, and Sheldon Solomon based upon Ernest Becker's ideas. Curse them for stealing my extraordinary thunder.

[2] *The Selfish Gene* by Richard Dawkins. Oxford University Press (1976)

at one point in our lives, it becomes an uncontrolled, misunderstood, whip-wielding monkey on our back. Our bodies, with those minute, physical, functional units that control us like a Master of Puppets, are merely vessels designed to perpetuate, to replicate those gene-engines. Love, romance, soppy music, Les Mis, perfumes, lipsticks, hair gels, gyms, are all just ways to get us to do this. We are being manipulated by nature to, in the words of Marvin Gaye, get it on.

> 'Researchers have found that the male brain is hardwired to seek out sex, even at the expense of a good meal, with specific neurons firing up to over-ride the desire to eat. Intriguingly, women do not have the same neurons, suggesting that sex for females comes secondary to sustenance.'[3]

Sex isn't, as we like to think, about fulfilling our own desires. It has nothing to do with us. It's about avoiding death – in the long term. The stigma that is still attached to those couples who don't have children is something that still sends conversations into a death-spiral and, in the back of many peoples' minds lays the question of their 'completeness'.

Food is the avoidance of death. From the moment we are born, we crave the nipple or the bottle. We are not trained to eat. We just know that we have to do it. If we don't get it, we cry like cats until someone shuts us the hell up.

We are, strangely, in our early years, dependent upon others to feed us because we are limited by our physical abilities but, once physically mature enough, put us on a desert island and we will fish and hunt and fight, even kill each other, for food.

Of course, society has turned this all into an elaborate dance through consumerism and the power of advertising, but it works; it works because, like sex, there is in us that innate need to gain the best food that we possibly can in the quickest possible

---

[3] www.telegraph.co.uk. 14 Oct 2015. This study was actually carried out on nematodes. I'm not sure how to interpret that.

time. *Lucozade Aids Recovery*; *A Mars A Day Helps You Work Rest And Play*; *I'm Lovin' It*; *Finger-Lickin' Good* (with a hyphen, it's so good); *Eat Fresh* – these slogans aren't accidental. Advertising is a deeply studied, psychologically intense field. Most adverts are prone to massive hyperbole, prey upon our weaknesses and fears and usually end up offering us something that we never actually thought we needed. I can remember, when I was a nurse, people brought Lucozade in to their relatives after bowel surgery. It was full of gas which, believe me, is not the thing that you want inside of you after losing fifteen feet of small bowel. A Mars a day didn't help you work rest and play – a balanced breakfast, lunch and dinner did that. A Mars a day just contributed to the world's obesity statistics. *Eat Fresh*, the slogan of Subway, is a clever little sausage of a slogan. Be honest, you are not going to go out looking for food that isn't fresh and yet the slogan offers us the one thing that our robotic inner selves all want. You are not going to join the crows and peck at sticky carrion at the roadside between passing cars. We know, instinctively, that food which is not fresh, will harm us. Look at how much food we throw away because it is past its sell-by date. We smell our food, not because it smells great, although it invariably does, but because there is something in us that tells us to smell it to check that it is safe to eat. Anybody sniffed the milk or the bread this morning? Anybody change the water in the bottle before they went on a bike ride? Who won't eat a bruised pear from the fruit bowl, apart from me? The expression, we eat with our eyes, is 100% true. Why do you think we fall for the patter that the chefs on TV sell us? Because they tart up the grub – if it pleases our eyes, we know that it will probably please our belly. We are suckers all. We are manipulated by those tiny motherboards within us that tell us, without us even realising it, that we must eat, when we must eat and what we must eat. It has all been twisted and turned by the politics of society, but the end result is the same – we avoid death.

And how does reading a book fit into all this? The sex and the food thing make sense, but how on earth can reading a book be related to death?

There are a couple of points in this theory's favour. One

is that books present knowledge. Knowledge enables us. It enables us to improve our minds and by doing so enables us to get better jobs, to know what to do when the bomb drops, to know how survive with Ray Mears in the snowy, death-trap forests of Canada. The more we know, the longer we are likely to live. Life is a process of learning.

Let's look at fishing. Put a stick in water, you probably won't catch a fish. Put something that wiggles and trails and leaves ripples on the water, the fish will see this as potential food and are more likely to be driven by their curiosity towards the object. Stick a worm or a fly on it and you're in Trout City. How about building a house? The story of the Three Little Pigs is a perfect example of the ability to survive through acquired knowledge. As a pastime, reading calms, diverts, reduces stress, takes us out of mundanity, fulfils our subconscious need to make ourselves better mates.

It doesn't matter that we read less and less from books (although it does to me), because we are able to get our information from so many sources nowadays. Knowledge is, quite literally, at our fingertips. Instead of taking months to research this book, I can reduce that time to weeks. If I can reduce that time to weeks I can, hopefully, make money from it more quickly, which means that there will be a guaranteed source of income for food and a reason for my mate to stay with me and propagate. She is less likely to die if she stays with me.

What about phobias? Well, that's obvious. Some people are afraid of water because they might drown. I'm afraid of spiders because I know, I just *know*, that a spider will bite me, I will die, it will cocoon me and then slowly suck the juices from my body until I am a dusty, desiccated shadow of myself.

It doesn't matter what you are afraid of, you are afraid because something tells you that the object that you fear, will do you harm – possibly fatally. We can, with methods such as aversion therapy, learn that those things we fear might not do the harm that we think they will, but the association between the object and the compromise to our safety has to be broken. The difficulty comes when we don't understand the source of the phobia; many of them are attached to deeply repressed events from childhood that we have somehow managed to shut away and yet still come to the fore when those fears are by some means brought to life.

The problem is that death is still a very taboo subject and is either greeted by silence or overt flippancy. We desensitise ourselves to it by, over the decades, increasing the violence in movies and in books. This gives us distance, tells us that it is something that happens to other people and makes us feel, albeit temporarily, better about our own mortality. The less we feel, the less affected we are by it.

When it comes home though, we are, more often than not, drowned in the awkward silence that accompanies it. We don't like to talk to colleagues about the recent death that they have endured because we don't want to upset them, we don't want to talk about something that we find difficult, that reminds us of our own mortality. Tears at funerals are very often not simply for the bereaved, but for those left behind, for what has been lost and for that which is yet to come. Funerals are terrifying because they hammer home reality.

In hospitals, we separate the living from the dying, not out of an act of kindness for the dying, but because we don't want to upset those who MRSA or medical incompetence have not yet killed. We shuffle the dying off into siderooms or we move them closer to the exit so that the mortuary trolley doesn't have to wobble its way through the ward like a faulty Tesco trolley when it comes to collect the body.

We don't speak ill of the dead because tomorrow it might be us and we don't want to think that someone might speak ill of us. We use euphemisms for death – asleep, deceased, late, lifeless, cold, departed, stiff, bereft of life (parrot, anyone?), bloodless, bought the farm, breathless, checked out, cut off (usually in his/her prime), defunct, demised, done for, expired, gone to meet his/her maker, gone to his/her reward, no more, offed, out of one's misery, passed away, pushing up daisies, resting in peace, sleeping.

I wrote a play called *Uncomfortably Numb* in which, I'm sure you'll be surprised to hear, death played an important part, and yet the main characters would not say 'dead', they would say 'demised', as if the use of the word would cause the fellow with the scythe to tap upon their shoulder, take them by the crook of the elbow and lead them away to whatever destiny awaited them.

So, why this book?

Partly for the reasons mentioned above. All that we do is dictated by death - birth, education, work, eating, sexing (it's still a thing), putting on the slap before we go out. It's like that game, Six Degrees of Separation, where everything finds a common thread back to the one thing that they all have in common. Death is our common thread.

I'm also writing about it because it's a fascinating subject. I dealt with it in its various forms when I was a nurse and was always intrigued by peoples' reactions; the false respect, the real reverence, the civilised routine of it, the cultural differences, the relatives who had been waiting poised for this moment to get hold of the deceased one's house and the relative who would have given their own life to have their loved one back.

I was also intrigued by my own reaction to the subject. In the end, I think it did for me. I don't think I was able to cope with the hypocrisy and the ritual and the constant reminder of my own mortality. There was very little more juddering than seeing a person's worldly possessions packed into an ASDA bag. It made me wonder if I would have much more to give away after I had 'fallen off my perch'.

I wrote about the first body I encountered, a young man of nineteen who had died in a road traffic accident, in *Condition of Life*:

> 'I stared at him. I couldn't help it. My eyes were drawn to him in the same way that people rubberneck a motorway accident... The thing about the dead is that their eyes are rarely shut. Their lids come to a rest about half to three quarters of the way to closed. Their lips are thin, their nose pinched, their skin waxy, but at any moment, they could wake up and, despite those

half-closed eyes, your instinct is that they are no more than asleep.
As I stared at him, I realised that there was something missing. This person in front of me was no longer meeting my expectations. His chest did not rise. I could not hear breaths. There was no movement in his eyes, no ruddiness to his young cheeks, no shine to his dark hair. Nothing living is ever completely still and yet the only thing he had was stillness. He was empty. Something had moved out. It is too easy to say that it was life that had gone, but it was also too difficult to grasp that concept. At that moment, I could not define life. I simply could not relate the lad in front of me to any concept that I could understand.'

It is something that never leaves you. A lot of the nurses I worked with remembered their first body. It was like the first blood of hunting – you were marked, changed for ever.

It is also an important subject and worthy of some sort of analysis or gathering of info in one place, not only in the philosophical parpage that has been the whim of the last few pages, but also because, statistically, it's just amazing. I will make you glad that you were born where you were born because the statistics for those living elsewhere are, thankfully, shocking. It's good to feel the warm comfort of someone else being statistically more likely to die before they are fifty than you are. I will look at the media's worrying attraction to death and how we lap it up like dehydrated dogs on a hot summer patio. I will look at diseases, murder, suicide, the gods, the decay of the body and ghosts and mediums. I will cover death from every angle except, with any luck, one.

Well, there you go. That's my theory. Life/death, yin/yang and all that. I'm sure someone else has come up with some similar theory[1]; there are many, far more intelligent people out there than me, but I haven't seen it and, believe me, I have looked for it.

Enjoy. It might give you something to look forward to somewhere down the, hopefully, very distant, line.

PART 1

WORLD DEATH RATES AND CAUSES

The population of the world is, give or take a couple, 7.5 billion. In 1900, it was 1.6 billion. In 1750, it was 700 million. In 1300 it was about 400 million. Once upon a time, it was 2. In 2050, it is anticipated to be 9.3 billion[4].

To give this a proportion we can relate to, in London, in the year 1, there were a few farmers, perhaps 50-100. By 1300, it had a population of 80,000-100,000, by 1500, 550,000-600,000. In 1901 it had grown to 6,506,954 and by 2015 it had risen to 8,615,246[5][6][7][8][9][10]. It is expected to be 13 million by 2050[11].

These explosions happened all over Britain[12] - Liverpool, Manchester, Sheffield - as the Industrial Revolution ground into top gear. What were once villages became smoke-belching, polluted, yellow, sulphur-fogged, overpopulated paeans to commerce.

TOWN	POP 1750	P0P 1861
LONDON	675,000	2,804,000
BRISTOL	45,000	130,334
BIRMINGHAM	24,000	296,000
LIVERPOOL	22,000	443,900
MANCHESTER	18,000	338,300
LEEDS	16,000	207,200
SHEFFIELD	12,500	185,200

---

4 www.ecology.com/

5 www.londononline.co.uk. The Agrarian History of England and Wales: Volume 4, Agricultural Markets and Trade, 1500-1750. Cambridge University Press.

6 www.demographia.com

7 Major Cities in the Middle Ages Tellier, L.N. (2009)

8 Urban World History: An Economic and Geographical Perspective. Presses de l'Universite du Quebec. Thirsk, J.; Chartres, J. (1990)

9 Greater London, Inner London Population & Density History. demographia.com.

10 https://en.wikipedia.org/wiki/History_of_London#Population

11 www.standard.co.uk

12 Table: Populations in major towns pre- and post-industrial revolution. The Census, 1801-1901: Statistical Reports. The National Archives

Up to 20.30 today, 3 December 2018, there have been over 330,000 births according to Worldometers. According to the same website, there have been about 138,000 deaths. That is over twice as many people born as dying. The world's population is already 7.6 billion and growing, although the live US census site has the population at 7.5 billion. Either way, it's a big day for the supermarkets. There is one birth every eight seconds and one death every eleven seconds. That is a net gain of one person every 14 seconds, according to the US government census on 3 December 2018. Up to this date, this year, the world population has grown by 75.5 million. That is made up of 130 million births and 54.5 million deaths. That is ridiculous. It's no wonder that we are running out of homes, fuel, power and God knows what else.

It makes me wonder if the world is slowly sinking in space. Maybe one day, like some errant snooker ball, we will gently bounce against the soft cushion of the universe and come to a stop, perfectly lined up with the yellow two-point sun, as if Stephen Hendry himself had gently tapped us across the black baize and left us in perfect alignment for our next shot at evolution.

As a species we are vulnerable. There are over two hundred different types of cancer. There are over fifty eye diseases. The heart is vulnerable to coronary artery disease, arrhythmias, congestive heart failure, heart valve disease, heart muscle disease and congenital heart disease. Within each of these categories are other categories; arrhythmias include sinus or nodal bradycardia, atrial flutter, supraventricular tachycardia, ventricular extrasystole, ventricular tachycardia and ventricular fibrillation[13], to name but a few. The bowel is a positive battleground of conditions that can cause the individual to waste away – Crohn's disease, ulcerative colitis, irritable bowel syndrome, coeliac disease, diverticulitis, cancer, Hirschsprung Disease. These have always been here. Imagine life before toilets. In *The Time Traveller's Guide to Medieval England,* Ian Mortimer vividly describes early Exeter, where the

---

[13] Oxford Handbook of Clinical Medicine 2nd edn

local brook, 'shitbrook' he calls it, is a dumping ground second only to the outflow from Sizewell B nuclear power station. As a breeding ground for further disease in an antibiotic free land, Shitbrook will do nicely, thank you.

We have 206 bones in our body, each of which can break in a hypnotically grisly way - a greenstick fracture is an incomplete fracture in which the bone is bent, a transverse fracture where the broken piece of bone is at a right angle to where it might once have been, an oblique fracture where the break has a curved or sloped pattern, a comminuted fracture is where the bone breaks into several pieces, a buckled fracture, where the bone ends are driven piledriver-like into each other, a pathologic fracture secondary to a disease process, causing the bones to weaken, open fracture dislocation and a stress fracture. This is only a selection of the types of fracture. The list is not endless, but it would certainly keep you occupied if you wanted to try out each one.

The chart above, from www.who.int, shows the worldwide causes of death for 2012. This relates to nations of all financial and social background. Ischaemic heart disease, stroke, diabetes, road injury and hypertension have all increased since 2000.

THE TEN LEADING CAUSES OF DEATH IN THE WORLD IN 2012	
ISCHAEMIC HEART DISEASE	7.4 MILLION
STROKE	6.7 MILLION
CHRONIC OBSTRUCTIVE PULMONARY DISEASE	3.1 MILLION
LOWER RESPIRATORY TRACT INFECTION	3.1 MILLION
TRACHEAL, BRONCHUS, LUNG CANCERS	1.6 MILLION
HIV/AIDS	1.5 MILLION
DIARRHOEAL DISEASES	1.5 MILLION
DIABTETES	1.5 MILLION

ROAD INJURY	1.3 MILLION
HYPERTENSIVE HEART DISEASE	1.1 MILLION

In low-income countries, lower respiratory infections are the highest cause of death, followed by HIV/AIDS and diarrhoeal diseases. In middle income countries, ischaemic heart disease is top, followed by stroke and then lower respiratory tract infections. It is similar in high income countries.

TOP TEN CAUSES OF DEATH IN LOW INCOME COUNTRIES PER 100,000 OF POPULATION 2012[14]	
LOWER RESPIRATORY INFECTIONS	91
HIV/AIDS	65
DIAORRHOEAL DISEASES	53
STROKE	52
ISCHAEMIC HEART DISEASE	39
MALARIA	35
PRE-TERM BIRTH COMPLICATIONS	33
TUBERCULOSIS	31
BIRTH ASPHYXIA AND TRAUMA	29
MALNUTRITION	27

TOP TEN CAUSES OF DEATH IN HIGH INCOME COUNTRIES PER 100,000 OF POPULATION 2012	
ISCHAEMIC HEART DISEASE	158
STROKE	95
TRACHEA/BRONCHUS/LUNG CANCERS	49

[14] www.who.int/mediacentre/factsheets/fs310/en/index1.html

DEMENTIA	42
CHRONIC OBSTRUCTIVE PULMONARY DISEASE	31
LOWER RESPIRATORY INFECTIONS	31
COLON AND RECTAL CANCERS	27
DIABETES	20
HEART DISEASE	20
BREAST CANCER	16

It is interesting to note these differences. Does a better way of life lead to increased heart disease and stroke due to obesity and a less active lifestyle? The media and the statistics would suggest so. So would EASO (European Association for the Study of Obesity):

> '•Worldwide obesity has nearly doubled since 1980.
> •In 2008, more than 1.4 billion adults, 20 and older, were overweight. Of these over 200 million men and nearly 300 million women were obese.
> •35% of adults aged 20 and over were overweight in 2008, and 11% were obese.
> •65% of the world's population live in countries where overweight and obesity kills more people than underweight.
> •Overweight and obesity are the fifth leading risk for global deaths. At least 2.8 million adults die each year as a result of being overweight or obese.
> •44% of the diabetes burden, 23% of the ischaemic heart disease burden and between 7% and 41% of certain cancer burdens are attributable to overweight and obesity.
> •More than 40 million children under the age of five were overweight in 2011[15].'

[15]www.easo.org/education-portal/obesity-facts-figures/ from World Health Organisation Fact sheet N°311

Their definition of obese is according to the BMI (Body Mass Index). This measures weight in relation to height. A BMI equal to or over 30 leads to a diagnosis of obesity.

When I was a nurse and required to check the BMI of patients at regular intervals, I have to admit that I found it to be a slightly dubious method. *I* was considered obese, according to our BMI chart in use at the time, when I weighed sixteen stone at 1.8 metres tall. There is a problem with proportionality which came up time and again. There was also a lack of accounting for the individual. However, I would say that, on a global scale, it is probably about as accurate as we will get with current methods and remains a good indicator of body mass in relation to health.

If you have seen the scenes on the spaceship in WALL·E, where humans are so fat that they can no longer walk, you will see where we are heading. I thought it was a documentary.

The high ranking of HIV/AIDS, diarrhoeal diseases and lower respiratory tract infections are indicators of lower education, lower income resulting in lowered ability to afford medicines (and perhaps more corrupt/inefficient government) and the lack of basic sanitation. We in the richer countries have educated ourselves to health. We have invested time and research and can afford to have a safety-net system where health care is available to all (although not for much longer, I don't think). It is interesting to note, and you can't have failed to have noticed it in the media, the rise in diabetes. We are eating ourselves to death. Diabetes does not even show up on the 2012 statistics for low-income countries. Preterm birth complications does show up (but not on the high-income chart), as does malnutrition.[16]

> 'In June 2001, the United Nations General Assembly declared HIV/AIDS to be "a global emergency". Member States agreed to…a comprehensive package of strategies for prevention and care, including:

---

[16] Charts from: www.who.int/mediacentre/factsheets/fs310/en/index1.html

- access to affordable condoms
- prompt treatment of other sexually transmitted infections (which increase the risk of infection with HIV)
- access to voluntary HIV testing and counselling
- prevention of mother-to-child transmission
-

Country By Wealth	**Life Expectation (years)**
1. China	75.2
2. United States	79.3
3. India	68.3
4. Japan	83.7
5. Germany	81
6. Russia	70.4
7. Brazil	73.6
8. Indonesia	70.6
9. United Kingdom	81.2
10. France[17]	82.4
Country by Lack of Wealth	**Life Expectation (years)**
1. DR of Congo	64.7
2. Liberia	62
3. Zimbabwe	58
4. Burundi	59.6
5. Eritrea	62.2
6. Cent African Rep	52.5
7. Niger	61.8
8. Sierra Leone	46
9. Malawi	58.3

[17] www.insidermonkey.com

10. Togo[18]	58.4

-
- promotion of advice and support to reduce HIV infection among intravenous drug users
- sexual health education in schools and the community
- improved access to care, support and treatment, including sustainable access to affordable supplies of medicines and diagnostics[19].

The implication that comes with the above statements is that these were not in place prior to the agreement. The statements are all about 'access' and 'promotion' and 'education', things not available to the so-called 'third-world'.

Life expectancy is also related to wealth and lifestyle (see table previous page), which includes education, availability to healthcare and living conditions.

In the wealth table, India and Indonesia are perhaps anomalies because of the vast, well-documented gap between rich and poor. China has a very different, healthier lifestyle to those in the west. Generally speaking though, the life expectancies shown do reflect a country's wealth.

It doesn't take too much to draw conclusions from the list of poorest countries, does it now?

The Japanese are the longest-lived with an average of 83.7 years. At the bottom of the scale is Sierra Leone, with an average of 46 years. That is a vast difference and is accountable to the above factors already mentioned. The UK is at number twenty, with 81.2 years. The US is at 31, with 79.3 years[20].

---

[18] www.therichest.com/rich-list/world/poorest-countries-in-the-world/

[19] www.who.int/immunization/topics/hiv/en/index1.html

[20] www.worldlifeexpectancy.com. Primary sources: WHO, World Bank, UNESCO, CIA and individual country databases for global health and causes of death. Published 2014.

In 2012, in England and Wales, the top deaths (presumably not as in favourite) for women was given as dementia and Alzheimer's, followed by heart disease and cerebrovascular disease. These were followed by chest infection, emphysema/bronchitis, lung cancer, breast cancer and bowel cancer. In men it was heart disease, lung cancer, dementia and Alzheimer's, emphysema/bronchitis, cerebrovascular disease, chest infections, prostate cancer, bowel cancer, lymphoid cancer and liver disease.

It's interesting to notice the differences and similarities. Men, it would seem are far more prone to cancers, perhaps because they smoke and drink more, but bowel cancer features highly for both men and women, perhaps due to the amount of red meat in the diet and the lack of roughage, perhaps because we are living longer (9 out of 10 cases are in the over 60s), because of increased alcohol intake, because the symptoms are more obvious than other types of cancer, because we delay treatment, putting those symptoms down to other causes, because obesity is high, because of familial histories and inactivity.

All of these stats are open to interpretation and will change with trends, with *fashion*. One would hope in the future to see less bowel and lung cancer as we are now more aware of the causes. Research into dementia is aggressive, with ongoing studies being carried out (my father has been in one for years) and more and more articles cropping up in the media on a daily basis.

The problem is that we are now being bombarded by facts and figures. In the space of a week, I was told by the BBC that drinking wine was both good and bad for me.

I am inclined to think that our knowledge is very much a double-edged sword. The more we find out about the ways in which we can go wrong, the more paranoid and confused we are liable to become.

In *Condition of life,* I wrote:

> 'What have I said before? 100% of the people who exercise will die. 100% of the people who don't exercise will die.

We are warned endlessly by the news about how fat we are becoming, how diabetes is increasing, how cancer is doing this and dementia is doing that.
Here's an idea:
STOP DOING RESEARCH. STOP IT!
Ignorance really, truly, most definitely is bliss. If you start digging in the cemetery, believe me, brothers and sisters, you'll find bones.
Have the number of people with diabetes or heart disease or cancer or God knows what other diseases there are, increased because we have found out so much more about them? Of course. Every new piece of research, every discovery, opens the door to both hope and disaster. If you want that hope, fine, but be prepared to take the disaster along with it… Have we got fat because we are eating too much shit? Yes. Did fat people exist many centuries ago? Yes. Is cancer a twentieth century disease? No. We used to call it by another name in the same way that we called tuberculosis consumption or tetanus lockjaw.

I love this from the Daily Mail:

'Cancer is a man-made disease fuelled by the excesses of modern life, a study of ancient remains has found. Tumours were rare until recent times when pollution and poor diet became issues, the review of mummies, fossils and classical literature found...Scientists found no signs of cancer in their extensive study of mummies apart from one isolated case. Michael Zimmerman…said: 'In an ancient society lacking surgical intervention, evidence of cancer should remain in all cases. The virtual absence of malignancies in mummies must be interpreted as indicating their rarity in antiquity, indicating that cancer-causing factors are limited to societies affected by modern industrialisation.'

www.dailymail.co.uk

There you go. It's your own bloody fault, the Daily Mail says so.
But then I found this (and please donate to the Help Americans Spell Foundation should you be in any way traumatised by this article):

The world's oldest documented case of cancer hails from ancient Egypt, in 1500 B.C. The details were recorded on a papyrus, documenting 8 cases of tumors [sic] occurring on the breast…In ancient Egypt, it was believed cancer was caused by the Gods.'.

www.cancer.about.com

What I am saying with all these facts and figures, is that we are made to be broken. We are born to die. We can do all we can to prevent that, we can research until the pasteurised, fat-free cows come home, but we will not succeed. The body is a finite thing. If we are lucky enough to have kids, our genes, those manipulative, head-strong bullets that pierce the veil of death, will live on, and that is the only way that we can even touch immortality.

There is an innate desire to live forever. We express it daily through our fears, our phobias, our road signs and the maternity wards. We watch TV serials because the assumption is that we will see the end. Coronation Street has been on air since Henry VIII died; if that's not hopeful, I don't know what is. We live almost to the point of denial of death. It happens to others, on the news, in the movies, even to our nearest and dearest, but once we have buried them, once the news has passed, we retreat into our cocoon of unreality and, despite the repressed kernel of truth that hides shyly behind our heart or rumbles in the diverticular pouches of our extensive guts, we keep telling ourselves that we will never die. Never.

I have some bad news for you…

PART 2

Disease

noun dis·ease \di-ˈzēz\

Simple definition of *disease*

: an illness that affects a person, animal, or plant : a condition that prevents the body or mind from working normally
: a problem that a person, group, organization, or society has and cannot stop[21]

[21] Source: Merriam-Webster's Learner's Dictionary

The world is rife with disease. We tend to think that, because we have hospitals and antibiotics and incredibly advanced surgery, that we will be fine, tickety-boo, on top of the world, ma. In our mind, I say again, we border on immortality. We think that whatever goes wrong can be put right by a visit to A&E or the GP or, if push comes to shove (it has to be a big shove though), the consultant in the clinic at the hospital.

In a way, I suppose we are taking tentative steps to that immortal (I have just typed 'immoral'; Oh, Mr Freud! What do you make of that mistake?) ideal. Life expectancy for males in 1841 was 40.2 years and for women it was 42.2 years. In 1901 it was 48.5 years for males and 52.4 years for females. Leaping to 1952, it had gone up to 66.4 years for males and 71.5 years for women[22]. These are still better than medieval times. In his book, *The Time Travellers Guide to Medieval England*, Ian Mortimer says:

> 'Yeomen in Worcestershire in the first half of the fourteenth century can, at the age of twenty, look forward to an average of twenty-eight more years life [48ish]; and their successors in the second half can expect another thirty-two years [52ish]…However, this bald figure means that half of all adults die before they reach fifty. And these are the *prosperous* members of…society. Poor peasants in the same area can expect to live for five or six years less...All these figures are for those who have…reached the age of twenty: half the population will die before this age. Life expectancy at birth can be as low as eighteen.'

Once again, wealth bought health. Eleanor of Aquitaine, mother of King John (booo! hisss!) and Richard the Lionheart (yaaaay!) lived until she was 92, but she was loaded and had Kings for husbands and sons.

Life expectancy has increased over the centuries as our

---

[22] http://visual.ons.gov.uk/how-has-life-expectancy-changed-over-time/

knowledge has increased. The majority of deaths on our early fifteenth century Tour De France (the one with Agincourt) were due to disease, such as dysentery and typhoid, including Henry V's death, but it was through this exceptionally tough learning process that we became more aware of our environment and the effect it had upon our bodies.

Florence Nightingale[23] washed her hands (sorry) of the ancient miasma theory and almost instinctively understood that there was something such as bacteria which caused infections in wounds and that, with the proper treatment, a substantial number of deaths could be prevented. Don't forget, Pasteur[24] and Koch[25] had not yet come up with the germ theory and it would be another century before antibiotics came into use.[26] Of the British, French and Russian casualties in the Crimean War of 1852-54, there were more than twice the number of deaths due to infection compared to those from a bullet.

> 'Nightingale believed that, by keeping patients well-fed, warm, comfortable, and above all clean...[The] Treatment of soldiers in Scutari provided an opportunity to validate this theory...To this task, Nightingale brought her skills as a nurse…prodigious managerial skills, an obsession with meticulous record keeping, and a deep faith in the Sanitarian movement…and although she presumably had no concept of bacteria or viruses, she clearly understood contagion…a clear relationship between the diseases killing her patients and the filth in which they lay, the air they breathed, the water they drank, and the food they

---

[23] Picture: By not listed [Public domain], via Wikimedia Commons

[24] Louis Pasteur discovered that germs cause disease. In 1861, he published his germ theory.

[25] Robert Koch discovered how to stain and grow bacteria in a Petri dish. He was thus able to find which bacteria caused which diseases - septicaemia (1878), Tuberculosis (1882) and cholera (1883). www.bbc.co.uk

[26] http://cid.oxfordjournals.org. That would be Penicillin in WWII.

> ate. In her words, "The 3 things which all but destroyed the army in Crimea were ignorance, incapacity, and useless rules".'[27]

There is a fashion now to belittle Florence and her work, but the determination of a woman in the nineteenth century to set things right in a world dominated by arrogant men (doctors and generals at that, heaven help us) and to set her knowledge, imagination and interpretive abilities gained *through her own experience* to work were simply astounding and courageous. She should be held up as an example of intelligence, stubborn self-belief and self-awareness.

There are today over thirty thousand known diseases. I say 'known' because who knows what science will find tomorrow? We'll probably find out that cigarettes are good for us, at least thirty a day mind, and that green tea is as detrimental to us as heroin laced with strychnine. Over seven thousand of these diseases can be considered 'rare diseases', a rare disease meaning that less than two thousand people suffer from it[28].

> 'Infectious diseases are still the major health problem of the majority of people inhabiting the earth. In the developing countries, the principal causes of death are infectious diseases…'new' diseases are being introduced by invasive diagnostic techniques, immunosuppressive therapies, changing cultural behaviour and sexual patterns.[29]'

The disease process itself needs several things to kick it off:

---

[27] Hamlin C, Sheard S. *Revolutions in public health: 1848* and Nightingale F *A contribution to the sanitary history of the British army during the late war with Russia.* London: Harrison and Sons; 1859.

[28] www.gesundheitsforschung-bmbf.de/en/131.php

[29] *The Textbook of Adult Nursing* – Brunner and Suddarth (1992)

1) A susceptible host
2) Exposure of the host to the disease
3) The right environment

That is all it takes really. It helps if the circumstances and the environment are right, for example, as in Henry V's army at the time of Agincourt and Harfleur that lived in filthy, unhygienic conditions and thus permitted the outbreak of dysentery or the refugee camps such as that which sprung up at Calais as a result of the unrest in the Middle East.

In diseases which are not infectious, such as cancer, there still needs to be something that stimulates the process, such as asbestos in mesothelioma or cigarettes in lung cancer. Unfortunately, as humans, despite all we know, we still go about blissfully (or willingly) unaware that we are constantly creating conditions wherein these diseases can flourish. 1 in 5 (19%) of adults in Great Britain smoke: 20% of men and 17% of women. Every year, around 96,000 people in the UK die from diseases caused by smoking. Tobacco smoke contains over 7,000 chemical compounds. These include carbon monoxide, arsenic, formaldehyde, cyanide, benzene, toluene and acrolein[30]. However, we rarely say, 'I'm just popping outside for a quick dose of arsenic and carbon monoxide'.

Disease then causes changes within the body. These changes are called *pathological*, because they are related to disease. Anything which changes the normality (whatever that is, pre-disease) of the body due to the disease process is pathological. A common and frequently misused phrase is 'pathological liar'. The lying would have to be due to a disease process, such as a brain tumour, which can lead to changes in personality, or mental illness.

Most diseases have an 'incubation period' for infectious diseases or a 'latency period' for non-infectious diseases[31]. Cancer might well be brewing asymptomatically for a very long time before any symptoms occur. The same can apply to infectious diseases, but

---

[30] Ash: *Facts at a Glance: Smoking Statistics* (June 2016)

[31] www.cdc.gov

generally, they tend to display themselves somewhat more quickly that the non-infectious ones. For example, chickenpox takes ten to twenty-one days to appear. Salmonella can take six to forty-eight hours. There are, of course, exceptions such as HIV/AIDS which can take one to fifteen years. If you were unlucky enough to be a watch dial painter, you might have used radium to give the hands that glow-in-the-dark fun look. It might take from eight to forty years for bone cancer to develop. The onset of symptoms marks the transition from subclinical to clinical disease[32].

However, screening methods available today, such as breast cancer screenings and blood tests for prostate cancer, can prevent (postpone) death by permitting earlier treatment. Relying upon treatment once symptoms have occurred will obviously be less effective in most cases.

Early disease was often treated with, by today's standards, ridiculously supernatural reverence.

> 'When the King of France asks the faculty of medicine at the University of Paris to explain the causes of the Great Plague of 1348-9, the worthy professors noted that the plague was due to
>
>> an important conjunction of the three higher planets in the sign of Aquarius, which, with other conjunctions and eclipses, is the cause of the pernicious corruption of the surrounding air, as well as a sign of mortality, famine and other catastrophes...the conjunction of Saturn and Jupiter brings about the death of peoples and the depopulation of kingdoms...the conjunction of Mars and Jupiter causes great pestilence in the air.'[33]

Sir Humphrey Appleby would have been proud of such

---

[32] www.cdc.gov

[33] *The Time Travellers Guide to Medieval England* - Ian Mortimer

gobbledegook. But it was all they had. This was a King of France in his court. Money was of no object when it came to royal comfort and safety, but scant scientific knowledge had to succumb, if not entirely to bullshit, then to some ethereal device.

I have to put this bit in, from the same source. It is so desperate and sad and, I shamefully admit, funny:

> 'If you suffer from quinsy (an abscess in the throat following untreated tonsillitis), the following remedy might be prescribed:
>
> > Take a fat cat, flay it well, and draw out the guts. Take the grease of a hedgehog, the fat of a bear, resins, fenugreek, sage, honeysuckle gum and virgin wax, and crumble this and stuff the cat with it. Then roast the cat and gather the dripping, and anoint the sufferer with it.'

I am a big fan of Ian Mortimer. His work is always engrossing and fabulously, impeccably, richly researched. Give it a go, but only after you have bought and read this, obviously.

Ignorance forced people to look for answers in the gods, to see it as punishment for the errant ways of humanity. It was believed that disease was carried in the 'miasma'. This propagated the idea that disease was spread through bad air. This is why doctors wore the 'bird' suits[34]. The 'beaks' were filled with herbal material to keep the plague or other diseases away. It wasn't until germ theory was introduced that the miasma theory was put aside.

There was also a strong belief in the four humours, as posited by Aristotle and Hippocrates (whence comes the Hippocratic Oath). These were yellow bile, black bile, blood and phlegm. These related, in order, to the natural elements, fire, earth, air and water.

---

[34] Picture of 'plague doctor' - By I. Columbina. Paul Fürst (1608–1666) was the publisher, and perhaps also the engraver. Public domain, via Wikimedia Commons

> 'Man's body has blood, phlegm, yellow bile and black bile. These make up his body and through them he feels illness or enjoys health. When all the humours are properly balanced and mingled, he feels the most perfect health. Illness occurs when one of the humours is in excess, or is reduced in amount, or is entirely missing from the body.[35]'

This was because there was a strong belief in the unity between patient and universe. This was the job of the doctors, to see that these four humours were balanced. An excess of one would be to the detriment of the other and it therefore had to be reduced. This is how we ended up with 'bleeding' where leeches were used to drain excess blood. Unfortunately, by this method, patients could often be 'bled dry'.

> 'One of the reasons why humoural theory continues to hold such sway is because it is so involved and complex. Its humerical harmony (the four elements, the four humours etc.) allows for endless refinement and invented complexity. If two men are identically injured - let us say, with a sword cut to the lower arm - a physician will treat them differently according to whether they are sanguine in temperament or melancholic. Alternatively, if two men are sick with the same disease they may receive wholly different treatments due to variations in the appearance of their urine. The physician will ask each man for a sample and judge it according to its colour, scent and cloudiness.
>
> Urine which is milky on the surface, dark at the bottom and clear in the middle is a sign of dropsy. But ruddy urine in a dropsical patient

35 Hippocrates, *'On the Constitution of Man'* (c.500BC)

> is a sign of death . . . urine which is red, like blood, is a sign of fever caused by too much blood; blood should be let immediately when the moon is passing through the sign of Gemini. Green urine, coming after red, is a sign of inflammation, a mortal sickness . . . saffron-coloured urine with thick, smelly and frothy substance is a sign of jaundice . . .'[36]

I do not wish anybody to think that the history of disease and medicine held no more depth than a farcical sitcom. Knowledge gained was logical. It made sense at the time and was, in that way, completely justifiable. It is only through empirical study that such things advance and this is what happened, but not overnight.

> '…Greek doctors at Alexandria in Egypt began to dissect bodies. Some even dissected the bodies of criminals who were still alive (vivisection). In this way the surgeon Herophilus [335–280 BC] realised that the brain, not the heart, controls the movement of the limbs, and Erasistratus [c. 304 – c. 250 BC] discovered that the blood moves through the veins (although he did not realise that it circulated[37])…Greek philosophers such as Thucydides realised that prayers were useless against illnesses...Hippocrates's…suggested that disease was caused by the environment. Thus the way was open for an entirely natural theory of the cause of disease.
>
> Based on the theory that natural matter comprised four basic elements, the Greek philosophers came

---

[36] *The Time Travellers Guide to Medieval England* - Ian Mortimer

[37] This honour belongs to William Harvey (1 April 1578 – 3 June 1657). He was the first to describe the systemic circulation and properties of blood being pumped to the brain and body *by the heart*. He used others' work as a basis for his, but he was fab, so I vote for him.

up with the idea that the human body consisted of the four humours, which had to be kept in balance.[38]'

I would like to look now at some of the diseases that have dominated the world, killed millions and, in many cases changed the course of our history. There is always an element of social change that comes with disease or its suppression.

The areas I will look at are:

- Plague
- Cholera
- Ebola
- HIV/AIDS

These diseases have been chosen for fairly obvious reasons – plague changed the social contract and went a long way to disposing of that feudal mentality that was so entrenched within society. From tragedy came enlightenment. Ebola has been in the news tremendously in the past couple of years, the impact of modernisation making it perhaps all the more deadly on a worldwide scale. HIV changed the way we live and think, forced us to control our baser instincts and even changed James Bond (for a short time) into a one-woman man.

We like to think that we have eradicated these diseases. We have become detached from them by time and physical distance, but they are still here, they always will be, either dormant or unnoticed by science until, one day, they erupt and cause worldwide panic like the Ebola virus did so recently.

---

[38] www.bbc.co.uk

The Plague

From KTVB.com, June 2016:

> BOISE -- Two pet cats in Idaho contracted the plague, and officials at the Central District Health Department say people need to take precautions in plague-impacted areas. Health officials say plague killed a cat in Mountain Home after it came into contact with ground squirrels. A pet cat in Clark County also recently became ill. "It's a bacterial disease that's very dangerous for people and pets especially cats and dogs," said Sarah Correll, an epidemiologist with the Central District Health Department. Dead ground squirrels in rural Southwest Idaho tested positive for plague earlier this year.

This is real[39]. It is from this year – on June 09, 2016. The twenty-first century. There is a common belief, hope, that we had left this behind in the fourteenth century or, at the very latest, the seventeenth century when, in 1665, plague once again ravaged the country but was, conveniently, burned away by the Great Cleanse of the Great Fire of London in 1666.

Time and distance. Out of sight, out of mind.

Yet, surprisingly, the plague is prevalent, endemic even, in many parts of the world.

Take the good old US of A, seeing as we have started there.

In October 2015, CNN reported that, according to the CDC (Centers for Disease Control), fifteen people had already been infected with bubonic plague that year; two in Arizona, four in Colorado, four in New Mexico, one in Oregon, one in California, one in Utah, one in Georgia and one in Michigan. The patients in Georgia and Michigan were infected in California and Colorado[40]. Four of those people had died. On 30 December, the Culver City Patch reported that the total was sixteen people,

[39] www.ktvb.com

[40] www.cbs4indy.com

higher than the average. According to the CDC, the death rate is 16% among patients who have been treated, and between 66% and 93% among those who are not treated.

The article stated that, on average, there were seven cases a year, but in 2006 there were seventeen. Plague occurs in rural and semi-rural areas of the United States, 'most commonly New Mexico, Arizona and Colorado. This year's cases have been reported in eight states.'

Now, I don't know how you, dear reader, feel about this, but I was kind of worried about rabies, never mind the plague.

Around the world in 2013 there were 783 cases of plague reported, including 126 deaths. It is endemic (permanent) in many countries in Africa, the former Soviet Union, the Americas and Asia[41]. In 2014, there were 482 cases reported in Madagascar. There were 81 deaths.

Signs and Symptoms

So how will you and I know if we have, through no fault of our own but for a couple of careless steps towards the red area of the map, caught the plague?

Well, thankfully, the documentation on this is pretty good. Unlike the rest of us, the CDC and the World Health Organisation did not bury their heads in the sand and kept a close tally on the spread of the disease.

It is surprisingly active worldwide

There are three types of plague:

1. Bubonic plague
2. Septicaemic plague
3. Pneumonic plague

They are not necessarily however, unconnected.

The **bubonic plague** is so called because of the buboes that appear in the groin, armpit and neck. I shall return to *Great Tales from English History* by Robert Lacey for a description:

[41] www.who.int

> 'Woe is the shilling in the armpit. It is seething, terrible, wherever it may come, a head that gives pain and causes a loud cry, a burden carried under the arms, a painful angry knob, a white lump. It is of the form of an apple, like the head of an onion, a small boil that spares no one. Great is its seething, like a burning cinder, a grievous thing of an ashy colour.'

There is an awful lot of seething in this contemporary account. Not that I blame them. According to Dictionary.com and Merriam-Webster, the word 'bubo' comes from the Greek for 'groin' and 'swelling', boubṓn, which seems spitefully appropriate - an interesting fact that you can use down the pub.

This is the commonest form of the plague and is caused by the bite of an infected rat. The bacteria that causes the plague is Yersinia Pestis (discovered in 1894 by Alexandre Yersin), which maintains its existence in a cycle involving rodents and their fleas. Through the bite, the bacteria gets into the lymphatic system, a sort of drainage system which runs in close proximity to the blood stream and picks up and disposes of infections. It then reaches things called lymph nodes, which are a sort of lobster pot, at junctions within the lymphatic system for catching nasty things, which then become inflamed. As the lymph node swells and becomes very tender, it becomes a bubo. These buboes can turn into large suppurating open sores. 'Victims are so overwhelmed that they're more or less poisoned to death as the bacilli gather in thick clots under the skin, where a passing flea might pick them up. Other grim side effects can include gangrene, erupting pus-filled glands, and lungs that literally dissolve'[42]. Other symptoms might include chills, malaise, a high fever, muscle cramps and seizures. Pain may well occur in the area before buboes appear. Gangrene of the extremities such as toes, fingers, lips and tip of the nose is a possibility. Interestingly, bubonic plague cannot be

---

[42] http://science.nationalgeographic.com/science/health-and-human-body/human-diseases/plague-article/

transmitted from person to person. The time between being infected and developing symptoms is typically two to seven days, but may be as short as one day for pneumonic plague[43].

Septicaemic plague, which spreads in the bloodstream, comes either via fleas or from contact with plague-infected body matter[44]. It is a (word of the day) zoonosis, that is a disease spread from animals to humans. It is essentially the stage where septicaemia, a suddenly newsworthy health item this year, sets in. Septicaemia is simply a widespread infection in the body. At this stage, the infection has reached the blood and is carried everywhere. It will eventually lead to organ failure and death if not treated. It is not unique to the plague. I have known patients with pancreatitis, chest infections and diabetes-induced gangrene succumb to septicaemia. It is a complication of disease. The patient might actually die before the symptoms appear and there are certainly records of people being fine at breakfast and dead by tea. You can't blame the cook every time.

Symptoms of septicaemic plague include abdominal pain, bleeding due to blood clotting problems, diarrhoea, fever, nausea and vomiting[45].

This is where **Pneumonic plague**, steps up to the plate. This is the most infectious of the three types of plague and is really an advanced stage of bubonic plague. It can be passed directly from person to person via airborne droplets which have been coughed from the lungs. The symptoms can appear suddenly, typically 2 - 3 days after exposure and include cough, difficulty breathing, fever, frothy, bloody sputum and pain in the chest when you breathe deeply. Pneumonic plague is 100% fatal. Treatment reduces this figure to about 50%.

There was a total of 6,622 plague deaths reported to WHO between 1954 and 1998[46]. None have been reported in

---

[43] www.nytimes.com September 10 2016

[44] http://science.nationalgeographic.com/science/health-and-human-body/human-diseases/plague-article

[45] www.nytimes.com September 10 2016

[46] www.who.int/csr/resources/publications/plague/asiaTable_3_1.pdf

Europe. Notice the word 'reported'. Many of the cases were in countries where reporting is either not a priority or just not possible. A cynical person might even suggest that advertising such things might affect the tourist trade and are better therefore not reported. I am not that cynical. Some people are though. But not me. A rough calculation on the research I have done of the notable outbreaks of plague in history accounts for about 85,000,000 deaths, but I think this is probably a vast underestimate when the above reasons are taken into account.

Social Implications of Plague

There was a certain structure to society in the Middle Ages. I am sure you'll be surprised to hear that this structure favoured the wealthy, not only in financial terms, but health-wise, dress-wise, food-wise, accommodation-wise and in any other wise you can think of.

Society had a strict social structure that went something like this:

Dukes, earls, barons, knights, esquires and gentlemen with £200 or more income from land, esquires and gentlemen with £100 income from land, franklins/yeomen, freemen, villeins (unfree labourers), domestic servants and then beggars[47].

This was strictly adhered to, right down to worship, food and dress. Lords worth £1000 a year (a vast amount, equivalent to £447,680 in 2005) could wear what they wanted. Knights with land worth 400 marks [£96,917.33 at 2005 values] could not wear 'weasel fur, ermine or clothing of precious stones' other than the jewels in a woman's hair. At the other end of the scale, the yeoman could wear fabric worth no more than 40 shillings (£895 in 2005) and no jewels, gold, silver, embroidery, enamelware or silk. No fur except lamb, rabbit, cat or fox; women were not to wear a silk veil. The likes of me, just above beggar, could wear no cloth but blanket and russet and belts of linen.

There was a clear demarcation in status.

---

[47] *The Time Travellers Guide to Medieval England* - Ian Mortimer

What the Great plague of 1348-9 and the subsequent plagues of 1361, 1368, 1374, 1379 and 1390 did was to go some way towards equalising society. It gave the poor power.

One half to two thirds of Europe died. By 1400 about half of all those born over the previous seventy years had died[48]. In England, employees simply did not turn up to work. They were dead. In these days of zero contracts, we would require a doctor's note, but back then, the doctors were dead too. Half to one-third of the workforce had gone. What this did was create a demand, made a man who was willing and able to work a valuable asset. People left their masters and went to work elsewhere. Workers pushed their prices up. Goods became scarce, prices went up, inflation spiralled. Edward III passed a law, the Ordinance of Labourers which fixed wages, prices and required all those under sixty to work. People ignored it. The Ordinance of Labourers was reinforced by the Statute of Labourers in 1351:

> 'It was lately ordained by our lord king, with the assent of the prelates, nobles and others of his council against the malice of employees, who were idle and were not willing to take employment after the pestilence unless for outrageous wages, that such employees, both men and women, should be obliged to take employment for the salary and wages accustomed to be paid in the place where they were working in the 20th year of the king's reign, or five or six years earlier; and that if the same employees refused to accept employment in such a manner they should be punished by imprisonment, as is more clearly contained in the said ordinance.'

People ignored this too.

'Skilled manpower was so short that…a carpenter who had been paid 3d. [£6.83] in 1346 was being

---

48 *The Time Travellers Guide to Medieval England* - Ian Mortimer

> paid 5d. [£11.39] by 1367, his mate had shot up from 1½d. [£2.98] to 4d. [£5.97], and most other workmen had added at least a penny to their wages.[49]'

However, with rising income comes inflation. The price of wheat went up 150% between 1348 and 1351. In France, wheat prices went up by 300%. In reality, wages probably fell when compared to the enormous rates of inflation imposed by the lack of labour and the concomitant lack of produce.

> 'After 1348-9, money-wages rose...But the cost of living grew…during the plague "grain rotted in the fields for want of men to harvest it" ...There was insufficient labour to plant the next year's crop. Food supply was squeezed… things changed in 1375. There was a bumper harvest. Crop prices plummeted. Yet after another major plague epidemic between 1368-71, the cost of labour remained high. Farming revenues fell, but costs did not budge. Feudalism was in a tricky spot. Manorial records of the late 14th century are full of despondent reports. Some lords abandoned their holdings. Others were forced to surrender to tenants on almost any terms the peasants cared to offer. Workers were paying lower rents—and had fewer obligations to their lord. Labour services faded out, to be replaced by purely monetary arrangements between employers and employees. These arrangements became customary, and led to the dissolution of feudalism by the 16th century.[50]'

The plague was a great leveller. Even the king lost his

---

49 www.bbc.co.uk/history

50 www.economist.com/blogs/freeexchange/2013/10/economic-history-1 Oct 21st 2013

daughter, Joan, when she was just thirteen. Society was rocking upon its caste-driven foundations.

It took many years for us to get where we are today, but with the Black Death, the waves of revolution began to lap at and erode the hierarchical cliffs. Wat Tyler's rebellion of 1381 might never have taken place had it not been for the consequences of the plague and the courage and confidence that it instilled in the indentured man. Without Wat Tyler, we might not have had the removal of Richard II, the Wars of the Roses, Jack Cade's rebellion, the English Civil War. Rebellion is infectious, from the lowest to the highest, and lives long in society's genes. All those events chipped at the colossus of social hierarchy under whose legs we peeped and eventually brought us the liberties we so treasure and take for granted today.

All because of a flea.

> 'The Black Death had horrific social effects. And the plague recurred sporadically until the 19th century. But by forcing the creation of monetised labour markets, as well as encouraging innovation and exploration, it spurred a weakened Western Europe towards economic development. The West's cultural superiority may not have been behind its eventual meteoric economic growth; disease did it.[51]'

Treatment

At the time of the Black Death, as we have heard, stuffing a cat with herbs and cow fat, cooking it and anointing oneself with the juices was one of the more *interesting* solutions offered. I don't know if you got to eat the cat. It seems a waste, if not.

It was however a very spiritual age. It had to be. There were no antibiotics. Cleanliness as a route to staying alive was not

---

[51] www.economist.com/blogs/freeexchange/2013/10/economic-history-1 Oct 21st 2013

expounded. King John took three baths a year and even that was a bit too metrosexual for their tastes. Humanity had absolutely no idea of bacteria, the transmission of disease, even the way our bodies worked. Eight hundred years later, to be honest, we only know a bit more.

So it was left very much to superstition, God and, frankly, wild lunacy involving cats.

Disease was carried in the miasma, the bad air, as I have stated previously. This is why doctors wore the bird suits. Herbs and nice smells were hidden in the beaks. Disease could also get in through the pores of the skin. Isolation and rapid burial were the only real options.

Praying was the favoured method. People turned to God.

> 'Although Edward did not know it, His prayers probably came a few days too late. By October, Dorset was overwhelmed with suffering. Towns which traded with the area ceased to welcome travellers. Those with the wherewithal, aware of the terrible mortality across the Channel, removed themselves to their most isolated estates, and stayed there. All England was plunged into fear.[52]'

There was a large amount of self-examination (not in the armpit sense) and righteous condemnation going on:

> 'Divine retribution also figured heavily in England's diagnosis, with the monkish chroniclers pointing a finger at the debauchery that went on at the royal tournaments. Between jousts the crowds would be entertained by female cheerleaders, some of them dressed like men in tight-fitting costumes that showed off their figures. 'We are not constant in faith,' complained Thomas Brinton…in the 1370s and '80s. 'We are not honourable in the eyes of the

[52] Ian Mortimer – *The Perfect King* (2008)

world.'[53]'

Prior to the introduction of antibiotics, not much changed from the methods employed as stated below by C N Trueman in *Cures for the Plague*[54]:

> '…brimstone 'burnt plentiful' was recommended for a cure for the bad air that caused the plague. Those employed in the collection of bodies frequently smoked tobacco to avoid catching the plague. "For personal disinfections nothing enjoyed such favour as tobacco…even children were made to light up..."…When money was used in day-to-day transactions in shops or at market, it was placed in a bowl of vinegar rather than being handed over to the recipient. At markets, meat was not handed over by hand…but by a joint being attached to a hook.
>
> The wearing of lucky charms was also common – and recommended by doctors…Dr George Thomson wore a dead toad around his neck. The Church had a more basic way of protecting yourself against the plague. It recommended prayer and then more prayer.'

Nowadays, antibiotics are used – streptomycin, Levofloxacin, Gentamicin, Ciprofloxacin, Doxycycline, Moxifloxacin (which has to be the best drug name ever) and Chloramphenicol. As can be seen by the worldwide death figures previously mentioned, either they don't always work or people do not have quick enough access to them to prevent death. With the evolution through mutation of bacteria and viruses, alternatives to antibiotics are being sought.

No doubt, the people who couldn't afford or get to the

[53] Robert Lacy – *Great Tales from English History* (2007)

[54] www.historylearningsite.co.uk/stuart-england/cures-for-the-plague/

antibiotics in time won't get that either.

Cholera

A brief History of Cholera (with a nod of the head to *Game of Thrones*)

You know nothing John Snow!

Well, as it turns out, he did. Quite a lot actually.

I must confess though that it's not the same John Snow as was murdered and brought back to life by a sexy witchy woman in *Game of Thrones*.

This John Snow is the one that, in 1854, identified the source of a cholera outbreak in Soho in London. He identified a water pump in Broad Street, had the handle of the pump removed and, from then on, the cases of cholera diminished.

What the process did was confirm for Snow that cholera was a water-borne disease. There had been over six hundred deaths already that year in that area alone[55]. That is staggering. Can you imagine the panic if a disease wiped out the occupants of a street near you? Worse still, if you don't know where the disease came from or the cause. The miasma theory, as was popular with the plague, was still at this stage being used to explain many diseases.

> 'At the time, it was assumed that cholera was airborne. However, Snow did not accept this 'miasma' (bad air) theory, arguing that in fact [the disease] entered the body through the mouth. He published his ideas in an essay 'On the Mode of Communication of Cholera' in 1849.'[58]

There was another epidemic in 1866, this time in the East End of London. The areas in question where supplied by a single water source, the Old Ford Reservoir, the waters of which were 'grossly polluted'.

It was from his conclusion that a lack of sanitation, specifically the lack of a clean water supply, was the causative factor in the outbreak and spread of cholera.

---

[55] *The English: A Social History 1066-1945* – Christopher Hibbert (1994)

The disease actually came to Britain, from the Ganges delta in India[56], the first recorded instance being in 1563 in an Indian medical report[57], long before John Snow became involved. It first came to England in 1831 (via Sunderland, of all places) and hit London (obviously, nothing matters until it affects London) in 1832. Much like the plague, once a disease is in situ, it doesn't take long for it to skip freely down the 19th century equivalent of the M1. That is not quite as flippant a statement as you might think. The Industrial Revolution, which called for the rapid transport of goods and people, both within the UK and across the world, helped to carry the disease. There were further outbreaks in 1841, 1854 and 1866. Between 1831 and 1866, there were at least 66,375 deaths due to cholera.

> 'The more widespread third pandemic of 1841-59 attacked the same regions as the second along with parts of south and central Europe. Subsequently, there was another massive outbreak from 1863-75 across the whole of Europe, large parts of north-eastern, South and Central America, Africa, China, Japan and Southeast Asia. The world continued to suffer the effects of cholera with a fifth pandemic in many parts of continental Europe, the whole of the North African coast and various areas in Asia and the Americas from 1881-96. London was to escape the ravages of cholera during this pandemic because its water supply had been transformed by the building of Joseph Bazalgette's sewage system. It was only when the Europe's other industrialized cities followed London's lead that Europe avoided further pandemics…'[58]

It did however continue to storm across Europe like

---

[56] www.who.int/mediacentre/factsheets/fs107/en/

[57] www.choleraandthethames.co.uk/cholera-in-london/origins-of-cholera/

[58] www.choleraandthethames.co.uk/cholera-in-london/origins-of-cholera/

Hitler on speed, with a *twelve-year* outbreak from 1863-75. From there it became a worldwide problem, not so much for Europe, but for Asia and the Americas and Africa it became and still is, a large contributory factor to death rates.

Today, cholera remains endemic in many parts of the world. In 2013, 47 countries reported a total of 129,064 cases (47% less than 2012), which included 2102 deaths[59]. Depending on the source, it is estimated that there are between 1.4 million and 5 million cases a year, with between 28,000 and 120,000 deaths a year.[60 60 64].

In 2014, cholera was reported in Cameroon, Democratic Republic of the Congo, Ghana, Niger, Nigeria and South Sudan, Cuba, the Dominican Republic, Haiti, Afghanistan, the Philippines, Nepal, India, Ethiopia, Tanzania, Kenya and Bangladesh[61 62]. It remains endemic in fifty countries (sixty-nine according to Stop Cholera in March 2015).

Once again, we have to look to poverty as a culprit, not just the *Vibrio Cholerae* bacteria. The countries mentioned above suffer from social inequality, poor income, corrupt government and poor recording of cases.

Signs and Symptoms

As has already been said, cholera is caused by the *Vibrio Cholerae*[63] bacteria – humans are the only known natural host, which reproduce in the intestinal tract, where the toxin that it produces causes the diarrhoea characteristic of this cholera[64]. The source of the cholera is usually in the faeces of the victim and, in

---

59 World Health Organisation - Weekly epidemiological record 1st August 2014 PDF

60 www.who.int/mediacentre/factsheets/fs107/en/

61 wwwnc.cdc.gov/travel/yellowbook/2016/infectious-diseases-related-to-travel/cholera

62 Stop Cholera: An Updated Estimate of the Global Burden of Cholera in Endemic Countries Updated March 27, 2015

63 Image: http://remf.dartmouth.edu/images/bacteriaSEM/source/1.html

64 www.britannica.com/science/bacteria/Bacteria-in-medicine#ref955503

poor sanitary areas, is therefore easily transmitted. If food or water is contaminated, the disease spreads.

The symptoms are:

- watery diarrhoea, sometimes described as "rice-water stools"
- vomiting
- rapid heart rate
- loss of skin elasticity
- dry mucous membranes
- low blood pressure
- thirst
- muscle cramps
- restlessness or irritability[65].

The symptoms usually take between a few hours and 5 days to appear and about 1 in 10 will display the symptoms mentioned above. The big danger is the diarrhoea. This leads to extreme fluid loss, dehydration and electrolyte imbalance. Renal failure, shock and death is a possible outcome.

Treatment and Prevention

When I was young, my father worked for the civil service and was posted overseas to Mauritius.

As was usual with people who had lived their entire lives in the Home Counties but for the occasional trip to the Isle of Wight, it sent a shiver of panic down our colonial spines, mostly because we knew that we were all going to die of a terrible disease, probably caused by street food, faeces-stained water or revolution.

The advice was to wash all food and only drink water out of a bottle. We all did that and none of us died.

There was sense to this, of course. We were, in 1976, when Mauritius was just connecting with tourism and the rest of

[65] http://www.cdc.gov/cholera/general/index.html

the world, vulnerable to disease, particularly with our soft southern immune systems. It was in Mauritius that, as a teenager I discovered alcohol. This was *sans* disease, obviously. I believe that this saved my life and happily continue to prove this to this day.

The advice we were given, which still stands today, is this:

- Drink only bottled, boiled, or chemically treated water and bottled or canned carbonated beverages. When using bottled drinks, make sure that the seal has not been broken.
- To disinfect your own water: boil for 1 minute or filter the water and add 2 drops of household bleach or ½ an iodine tablet per litre of water.
- Avoid tap water, fountain drinks, and ice cubes.
- Wash your hands often with soap and clean water.
- If no water and soap are available, use an alcohol-based hand cleaner (with at least 60% alcohol).
- Clean your hands especially before you eat or prepare food and after using the bathroom.
- Use bottled, boiled, or chemically treated water to wash dishes, brush your teeth, wash and prepare food, or make ice.
- Eat foods that are packaged or that are freshly cooked and served hot.
- Do not eat raw and undercooked meats and seafood or unpeeled fruits and vegetables.
- Dispose of faeces in a sanitary manner to prevent contamination of water and food sources[66]

I have to confess, I did eat a shitload of street food. It was gorgeous. I discovered it almost on the first day there. A street vendor around the corner from the hotel sold what were called Gateaux Piments – chili cakes – which were basically deep-fried

---

[66] www.cdc.gov/cholera/general/index.html

bundles of dough stuffed with chili and *Vibrio Cholerae.* I often spent my bus money on the food that the vendors sold in the school at lunchtime. I didn't die. My mother would have been so mad at me if I had.

So, prevention is better than cure, except in teenagers who drink too much and who are clearly immortal.

There are two cholera vaccines available, Dukoral (Swedish) and ShanChol (Indian), but these are not available in the USA. The CDC says that 'vaccines offer incomplete protection', while WHO says, 'the use of this vaccine is not recommended by WHO, mainly because of its limited efficacy and short duration of protection.' which seems a little bizarre on the basis that some vaccine/protection is better than none, but what do I know? Does cost effectiveness intrude upon their usage? Perhaps.

> 'The most recent and comprehensive cost-effectiveness analysis of cholera vaccination was conducted by the Diseases of the Most Impoverished Program…The results show that across [four] sites, only the programme for children aged 1–14 years in Kolkata would be cost effective (defined as the cost/ disability-adjusted life years [DALYs] averted being <3 times the gross domestic product per capita). However, when an immunological herd effect of cholera vaccination is taken into account, all programmes in all 4 countries would be cost effective, and programmes for children would be very cost effective in Beira and Kolkata (defined as cost/DALY averted being <gross domestic product per capita) and cost effective in Matlab.[69]'

If you go to:

http://www.idcostcalc.org/contents/cholera/cost-model.html#refs

you will see that we even have equations to weigh up the cost of a life.

Certainly, the use of Dukoral in their report *Weekly Epidemiological Record of* 26 March 2010, suggests effectiveness, but demands continuous rigorous monitoring and regular vaccinations.

With Shanchol and mORCVAX vaccines:

> 'The protective efficacy of the vaccine for all ages after 2 doses was 66%, and similar results were obtained in children aged 1–5 years and adults. The overall effectiveness of this vaccine 3–5 years after vaccination was 50%.[67]'

This book however is not about high horses and opinions, so I shall climb down and move onto alternative stuffs.

The Social Implications of Cholera

However, here I can talk about money, because it's relevant. Hear the hooves of the grumbling man on his nag.

> 'In 2007, various countries around the world notified 178,677 cases of cholera and 4033 cholera deaths to the World Health Organization (WHO). About 62% of those cases and 56.7% of deaths were reported from the WHO African Region alone. To date, no study has been undertaken in the Region to estimate the **economic burden** of cholera for use in advocacy for its prevention and control. The objective of this study was to estimate the direct and indirect cost of cholera in the WHO African Region.[68]'

---

[67] World Health Organisation - Weekly epidemiological record 26 March 2010

[68] http://bmcinthealthhumrights.biomedcentral.com/articles/10.1186/1472-698X-9-8

Funnily enough, I have come across a study that delves into the costs to the economy of cholera. It is only based on the African continent, but nonetheless gives a good idea of the costs involved. It comes from *BMC International Health and Human Rights* and is based on 2005 figures. They are also quite bare.

In 2005 there were 125018 cases of cholera in the African regions. They match life expectancy with costs

:

Life Expectancy	Total Economic Loss
40 years	US$39 million
53 years	US$ 53.2 million
73 years	US$64.2 million

What does this rather small amount of information tell us? Basically, the longer people live, the more expensive it is to keep them alive. This is a good argument for prevention over cure.

The figures for 2006 with 302,564 cases were:

Life Expectancy	Total Economic Loss
40 years	US$91.9 million
53 years	US$128.1 million
73 years	US$156 million

The main driver of the costs is the premature deaths of young children.

There are of course caveats that come with such a study – that the WHO guidelines are followed, with the concomitant standards of treatment and reporting, the assumption that those who died will have lost future earnings, which in turn affects GNP (gross national product) and GNI (gross national income), cost of diagnosis might have been overestimated in the study as it was assumed that all cases received a diagnostic test, the effects upon tourism and exports:

'...that leads to losses of revenues for the tourism

industry, unemployment of people whose livelihood is dependent on tourism, and reduction in tax revenues for the governments. In addition, usually, there is an international embargo on export of commodities from countries experiencing cholera outbreaks. When the latter happens, it may adversely affect the foreign exchange flow into the affected countries, which is likely to have many other externality costs'[70].

'For example, the cholera outbreak in Peru in 1991 cost the country US$ 770 million due to food trade embargoes and adverse effects on tourism.'

www.who.int

So those are the financial costs.

Let's not forget too that there are those invisible costs; the psychological impact upon the affected country and the outside world, the impact upon the bereaved, the poverty that surrounds the death of a family member because an income is lost, status is lost, maybe homes are lost or food not bought because the main money-winner is lost, the mother or father, with all the psychological impact that has on a child, the loss of a sibling and the learning and support lost through that. It all, though clearly not incalculable, everything is itemised these days, impacts upon the welfare of the individual, their community and, ultimately, their country.

INTERMISSION

Time for a fun-filled break.

Some fun-filled facts

1. About 100-108 billion people have already died since humans began.
2. Doctors' sloppy handwriting kills more than 7,000 people annually[69].
   I can testify to this. I knew one doctor who barely spoke English and had the handwriting of a gerbil. His writing was so bad that people would gather from far and wide to run bets on what the next word was. He was eventually fired for the equivalent of insider trading – he knew the answers and stuck a patsy in the pool.
3. Death by vending machines - Two people are crushed to death every year in the United States alone by trying to tilt faulty vending machines[70]. If it's the same two people very time, they must be fed up with it by now.
4. Falling coconuts cause about 150 deaths every year[71] - There is a 250,000 - 1 chance of this happening to you. Falling from a height of 80 feet, they can build up an impact speed of 50 mph[72]. I'm okay. I'm in Barnsley.
5. Over 2500 left handed people a year are killed from using products made for right handed people[73] - Evidence is inconclusive actually, but it makes you think about how we define our world.
6. 12 high school and college football players die each year[74]

---

69 http://content.time.com/time/health/article/0,8599,1578074,00.html

70 www.theguardian.com/theguardian/2010/dec/07/things-likely-kill-than-shark

71 www.wnyyradio.com/news/25-shocking-things-more-likely-to-kill-you-than-a-shark/

72 www.mirror.co.uk 30 May 2008

73 www.anythinglefthanded.co.uk

74 www.nydailynews.com/life-style/health

- 243 American football deaths recorded between July 1990 and June 2010. Amazing. Cristiano Ronaldo dies every time someone goes near him[75].

7. An American (where else?) company will take your remains and turn them into a diamond which can be used by your loved ones:

> *'The LifeGem is a certified, high-quality diamond created from the ashes of your loved one as a memorial to their unique and wonderful life. The LifeGem diamond can also be made from the carbon in a lock of hair.*
>
> *A LifeGem diamond provides comfort and support when and where you need it. Embrace your loved one's memory day by day.*
>
> *The LifeGem is the most beautiful and timeless memorial available for creating a testimony to their unique life. It will provide a lasting memory that endures time. After all, a LifeGem is forever.*
>
> *Call today for more information or to request your FREE LifeGem shipping kit. It contains everything you need to securely return your order to our office[76].'*

I admit it, this appeals to me, the thought that someone has to be burdened by my weight in carats for the rest of their lives and then force the kids to do the same…and they can't sell it/me on Flog It!. They'd be roundly castigated by the viewing public.

8. People sell tapeworms for weight loss[77] - In 2013, an

---

[75] Chris Bradbury Annoyance Level 10 Fact File

[76] http://www.lifegem.com/

[77] www.tapewormmartin.weebly.com

Iowan woman told her doctor she had bought a tapeworm on the internet and ingested it in an attempt to lose weight. While tapeworms can cause anaemia and malnutrition, it can't absorb enough food to significantly affect weight[78]

9. A headache and inflammatory pain can be reduced by eating 20 tart cherries[79]- When I saw this, I had two thoughts. One, that this was ridiculous and two, that it was no good to me because I don't like fruit, but:

The People's Pharmacy

Tart Cherries Fight Pain and Inflammation Naturally

Sour or tart cherries (also known as pie cherries) can reduce inflammation and pain from many causes when consumed fresh, canned, dried or juiced.

Terry Graedon September 2, 2013

Tart cherry juice contains natural compounds that have anti-inflammatory activity and can ease pain (Osteoarthritis and Cartilage, Aug. 2013). Other readers have reported similar benefit. One wrote:

"I used tart cherries to cure a gout attack and it worked. The real news is that the pain from osteoarthritis of the hip joint also diminished.

"I've been able to reduce my use of Celebrex

[78] http://facts.randomhistory.com/human-parasites-facts.html

[79] www.muskurahat.us

by half and still have less pain."

Give me some co-codamol any day.

10. Nerve impulses travel at over 249 mph[80] - This is why I'm always tired.
11. If you took all the blood vessels out of an average child and laid them out in one line, the line would stretch over 60,000 miles. An adult's would be closer to 100,000 miles long[81].
12. There are seven miles of new blood vessels for every pound of fat gained[82] - When you gain a pound of fat, your body makes seven new miles of blood vessels. This is why, when I weighed nineteen stone, I couldn't breathe anymore and I had a rapid pounding in my chest. It was my heart trying to escape. Fortunately, if you lose a pound, your body will break down and re-absorb the now unnecessary vessels. This means that I have lost 161 miles of blood vessels due to my recent diet. Be impressed. Be very impressed!
13. It takes food seven seconds to go from the mouth to the stomach via the oesophagus[83] – unless you are a hungry dog, when it barely touches the side of the mouth and skips the oesophagus entirely.
14. 85% of the population can curl their tongue into a tube[15] – This is relevant in our family. I can. The wife cannot. She has a stubby, unfortunate tongue. Any reader who hasn't tried this simply because I have mentioned it, is not paying attention.
15. You can live without a significant amount of inner organs, such as the stomach, the spine, 75% of the liver, 80% of

---

[80] www.medindia.net

[81] www.fi.edu/heart/blood-vessels

[82]http://articles.mercola.com/sites/articles/archive/2012/09/19/10-amazing-human-body-facts.aspx

[83] www.funology.com

the intestines, one kidney, one lung and any productive organ. It wouldn't feel so great, and you'll have to be under medication, but it won't kill you[84] – **Don't try this at home.**

[84] http://interestingthings.info/facts/16-human-body-facts-heart-lungs-blood-vessels-and-senses.html

END
OF
INTERMISSION

Ebola

A brief History of Ebola

People thought that this was funny. They sent text jokes and made facetious comments about it in a loud, proud voice to their friends.

Now, I'll laugh at pretty much anything. I have the sense of humour of a sociopath. The misfortunes of others usually rubs my funny bone like a genie's lamp.

That is, essentially, comedy. The misfortunes of others.

Ebola[85], however, was never funny. It has lain in the background for years like a stalker waiting for the chance to invade our house and finally, a couple of years ago, it did just that.

The thing about it that stunned me, and many others, was how slow the world was to react, how open the borders remained, how lightly those 'in charge' treated the whole affair. Once it was realised how serious it was, of course, the world went Ebola-crazy and each country, each organisation, clamoured to be the one that was seen most on camera, queuing up for interviews with the logo of their organisation on their chest more apparent than their face. One can't help wonder whether, had the outbreak been in London or New York, it might have been taken more seriously.

They were, however, very nearly too late.

> 'The World Health Organization was first alerted to the current outbreak of Ebola virus disease on 23 March 2014, but it was not until 8 August, after a meeting of the International Health Regulations Emergency Committee [one must have a committee before causing panic], that it declared a public health emergency of international concern. This official declaration set into motion an international response to contain the outbreak. The international response has been called both too small and too slow, and this may have contributed

[85] Image: www.public-domain-image.com

to the ongoing spread of the disease.[86]'

Jeremy J. Farrar, M.D., Ph.D., and Peter Piot, M.D., Ph.D. said in their article *The Ebola Emergency – Immediate Action, Ongoing Strategy* in *The New England Medical Journey* of October 16, 2014:

> '...our response to such events remains slow, cumbersome, poorly funded, conservative, and ill prepared. We have been very lucky with the severe acute respiratory syndrome, H5N1 and H1N1 influenza, and possibly the Middle East respiratory syndrome coronavirus, but this Ebola epidemic shows what can happen when luck escapes us. With a different pathogen and a different transmission route, a similar crisis could strike in New York, Geneva, and Beijing as easily as this one has in West Africa.'

Ebola is a comparative newcomer in the world of disease. It wasn't discovered until 1976 when there were outbreaks in Sudan and the Democratic Republic of Congo[87]. According to WHO, more than 600 people were affected by this outbreak[88]. One of the outbreaks took place near the Ebola River[89], from which the disease takes its name.

Between 1976 and November 2014, there have been a total of 31,079 cases with the total number of deaths at 12,922.

---

[86] www.bmj.com *International donations to the Ebola virus outbreak: too little, too late?* 03 February 2015

[87] www.theguardian.com September 25 2014

[88] www.bbc.co.uk/news/world-africa-2875503

[89] Peter Piot, currently director of the London School of Hygiene and Tropical Medicine but then a young infectious disease expert, first encountered the virus and together with colleagues decided to name it after a river that flowed through the district.

That's a percentage of 41.5% deaths[90].

According to a WHO Situation Report of 10 June 2016, 'a total of 28,616 confirmed, probable and suspected cases have been reported in Guinea, Liberia and Sierra Leone, with 11,315 deaths', broken down as such:

4,809 Liberia
3,955 Sierra Leone
2,536 Guinea
8 Nigeria[91].

The figures in the report are from December 20, 2015 and apply to the most recent outbreak, since January 2014. That's a 39.5% death rate. There will, undoubtedly, have been some cases and deaths unreported.

Causes, Signs and Symptoms

> 'Nearly 20 years later, in 2005, researchers looking for the reservoir of Ebola sampled more than 1,000 small animals in the Central African nations of Gabon and the Republic of the Congo, which have also experienced outbreaks of Ebola. They tested 679 bats, 222 birds and 129 small terrestrial vertebrates.'[92]

Bats, though not quite yet conclusively proven, do seem to be the main culprits. A report on Livescience.com says that 'it's possible that these bats [hammer-headed bat, Franquet's epauletted fruit bat - it suggests a well-dressed, cultured beast - and the little collared fruit bat] were sources for the outbreak'. A report in the Guardian of September 25 2014 said:

---

[90] Figures from www.cdc.gov/vhf/ebola/outbreaks/2014-west-africa/distribution-map.html#areas. Maths by me.
[91] http://www.bbc.co.uk/news/world-africa-28755033
[92] www.livescience.com/47946-where-did-ebola-come-from.html

> 'One reason for the fear surrounding Ebola is that no one knows where the virus goes between outbreaks. Primates have long been known to harbour Marburg [strain of virus]. Both it and the Ebola virus have also been found in three species of fruit bat in and around Gabon, bat soup is a delicacy in the region.'

This report and others such as the aforementioned one and the CDC lean strongly towards the involvement of bats and other mammals as carriers. The CDC says:

> 'Ebola is introduced into the human population through close contact with the blood, secretions, organs or other bodily fluids of infected animals such as chimpanzees, gorillas, fruit bats, monkeys, forest antelope and porcupines found ill or dead or in the rainforest…There is no evidence that mosquitoes or other insects can transmit Ebola virus. Only a few species of mammals (e.g., humans, bats, monkeys, and apes) have shown the ability to become infected with and spread Ebola virus.'

There is also the warning that the virus can be spread through sexual contact, even survivors of the disease might still be carriers. It is recommended that survivors only resume sex after their semen has been tested as negative for the virus twice. They go on to say that a person should be isolated, and the authorities contacted, if they have the early symptoms of Ebola and have had contact with:

- blood or body fluids from a person sick with or who has died from Ebola
- objects that have been contaminated with

the blood or body fluids of a person sick with or who has died from Ebola

- infected fruit bats and primates (apes and monkeys) or
- semen from a man who has recovered from Ebola

The signs and symptoms, variously described by WHO, the CDC and the Mayo Clinic are:

- Fever
- Severe headache
- Joint and muscle aches
- Chills
- Weakness
- Nausea and vomiting
- Diarrhoea (may be bloody)
- Red eyes
- Raised rash
- Chest pain and cough
- Stomach pain
- Severe weight loss
- Bleeding, usually from the eyes, and bruising (people near death may bleed from other orifices, such as ears, nose and rectum) [hence the name Haemorrhagic Fever]
- Internal bleeding

The symptoms might take anywhere between 2 and 21 days to appear[93].

I'll leave this section with this rather touching story from the BBC[94]:

---

[93] Image: By Mikael Häggström [CC0], via Wikimedia Commons

[94] www.BBC.co.uk/news/world-africa-30199004

> ‘It was one unsafe burial that ended up leading directly to Sierra Leone's explosion of Ebola cases in the summer.
> The country's first diagnosed case, when a pregnant woman was admitted to a hospital in the Kenema district following a miscarriage on 24 May, infected no-one else.
> Identifying the source of her infection, however, illustrates how the virus entered the country.
> The woman had attended the recent funeral of a well-known traditional healer. The healer had treated Ebola patients flocking to seek her care across the border from Guinea's Gueckedou region, before dying herself.
> Health teams working in the region identified a further 13 women who caught the virus attending the same burial, starting a chain reaction of infections, deaths and more funerals.
> According to the WHO, "quick investigations by local health authorities suggested that participation in that funeral could be linked to as many as 365 Ebola deaths".
> From there Ebola spread to Sierra Leone's capital Freetown where overcrowded living conditions and fluid population movements helped it to spiral further out of control.’

Now, add cars, lorries, vans, ships and aeroplanes into the mix and you can see why there was an urgency (on my part, at least) to close the airports, ports and borders. I’m not sure we’ll ever know how close we came to an outbreak here in the UK, but recent news events about health workers failing to observe guidelines, let alone plain, simple bloody common-sense and consideration, should ring alarm bells among us all. More than 240 health care workers developed the disease and 120 died[95]. It calls for caution and sense.

Treatment and Prevention

There are currently (at the time of writing in 2016) no licensed Ebola vaccines but research is ongoing.

There is however a joint project between the Sierra Leone Ministry of Health and Sanitation, the CDC and the College of Medicine and Allied Health Sciences at the University of Sierra Leone, called STRIVE. This rather apt, if not slightly shaky acronym, stands for Sierra Leone Trial to Introduce a Vaccine against Ebola. It does appear to be in its early days, but has at least got to the stage where it is being trialled on humans in five districts of Sierra Leone[96].

Diagnosis is by the usual round of bodily fluid tests, from blood to semen. Treatment is at present symptomatic with isolation, fluid replacement, control of pyrexia, maintenance of oxygen levels and blood pressure and the treatment of infections and symptoms as they arise being among them.

> 'Recovery from Ebola depends on good supportive care and the patient's immune response. People who recover from Ebola infection develop antibodies that last for at least 10 years, possibly longer. It is not known if people who recover are immune for life or if they can become infected with a different species of Ebola. Some people who have recovered from Ebola have developed long-term complications, such as joint and vision problems.[97]'

The Social Implications of Ebola

---

95 www.who.int/mediacentre/news/ebola/25-august-2014/en/

96 http://www.cdc.gov/vhf/ebola/strive/qa.html

97 www.cdc.gov/vhf/ebola/treatment/index.html

In countries where there is already a poor standard of nourishment, hygiene and income (Sierra Leone already being the eighth poorest of the nations with the lowest life expectancy of 46 years), the disease serves only to intensify the physical and psychological suffering already in existence.

Liberia has been described as 'on the brink of collapse'[98], made worse by the 14 year long civil war that, for all intents and purposes, rendered it without infrastructure.

There has been, in all the affected countries, a loss of domestic output, a devastating shortage of food, lack of employment and therefore income, and the countries' financial reserves are being eaten away as they attempt to prop up a euphemistically and literally dying nation. Education has suffered, the elderly, the poor and the disabled or chronically ill have suffered more than most because of their innate or acquired vulnerabilities and those involved in healthcare or jobs such as the burial of the affected dead have, in the words of reliefweb.int, been 'stigmatised'. 'Health systems have collapsed and non-Ebola related mortality is increasing', says the same report.

The key finding in the reliefweb.int report *Ebola in Sierra Leone: The Impacts of the 'Ebola Virus Disease' On the Livelihoods of Rural Communities, Agricultural Production and Food Security,* displayed the effect upon the economy from the grass roots to the banking institutions:

> '1. The people's main sources of livelihood have changed due to the outbreak of the EVD (Ebola Virus Disease). 97% of the surveyed households indicate that their income has dropped between May and August 2014.
> 2. The EVD is a driver of migration. Half of the people who have left their communities within the past four months, did so because of the epidemic.
> 3. In the epicentres of the Ebola outbreak the food production is decreasing: 80% of the surveyed

---

[98] http://www.bbc.co.uk/news/world-africa-30199004

households expect lower returns than last year. By-laws have discouraged many farmers from harvesting their fields. 71% of the interviewed households struggle to find laborers for their farms.
4. The EVD limits the availability and increases the costs of food. Certain foods have become scarce. The price of rice has - in average - risen by 30% since May 2014.
5. The Ebola epidemic has effects on community members' access to financial services: For 77% of the interviewed market vendors the access to credits has decreased since May 2014. In 5% of the surveyed communities, banks and microfinance institutions have even halted operations.
6. Market prospects have deteriorated in the wake of the continuing spread of the EVD. Two thirds of the interviewed traders remark that the volume of traded commodities has dropped significantly.
7. Awareness among rural communities of the dangers of the EVD has increased considerably: 83% of the interviewed households claim to know details about the disease and its transmission.[99]'

There have been riots in several countries and, in a region of the world already divided by tribal and religious differences, this might be enough to spark further troubles or even reignite civil war.

It's not over yet.

[99] http://reliefweb.int/report/sierra-leone/ebola-sierra-leone-impacts-ebola-virus-disease-livelihoods-rural-communities-0

AIDS & HIV

A brief History of AIDS/HIV

You are about to hear the sound of hooves and huffing and puffing again as this weary old man climbs back onto his high horse.

What is causing this elevated equine outburst? These words – The Gay Plague. I can't stand it. It reeks of righteousness and brimstone and condemnation and superiority and judgment and fear.

The appearance of HIV and AIDS in the 80s brought out the worst in the human race. It made our ignorance shine like a yellow, infected beacon towards stupidity and prejudice and sent certain types of people into the arms of their fellow racists and ethnicists and bigots.

I remove my feet from the stirrups and climb down, with a light-footed flourish, from my horse. I give it some carrot and thank it for its patience.

The Human Immunodeficiency Virus was first recognised in 1981, although it can be traced back to as far as the first third of the 20th century.

> 'My laboratory has had a long-standing interest in elucidating the origins and evolution of human and simian immunodeficiency viruses, and in studying HIV/SIV gene function and disease mechanisms from an evolutionary perspective. Characterizing the evolutionary relationships of simian immunodeficiency viruses infecting different non-human primate species in sub-Saharan Africa, we found that Acquired Immunodeficiency Syndrome (AIDS)…was the result of cross-species infections of humans by lentiviruses **of primate origin** [my bold]. Specifically, we discovered that HIV-1 resulted from cross-species infections of SIVcpz and SIVgor naturally infecting chimpanzees…and gorillas… respectively, while HIV-2 resulted from transmissions of SIVsmm infecting sooty

> mangabeys.... Moreover, we found that these viruses had entered the human population on multiple occasions, although only one of these transfers had spawned the HIV-1 pandemic. As we now know from molecular clock analyses, the main group of HIV-1, which has afflicted more than 70 million people and caused more than 30 million deaths worldwide, was transmitted to humans in the first third of the 20th century.[100]'

Acquired Immune Deficiency Syndrome occurs as a result of the HIV retrovirus, which leaves the individual susceptible to opportunistic disease. The virus destroys a type of white blood cell called CD4 cells. These cells help in the body's fight against infection. This is where the 'Immune Deficiency' part of AIDS comes into play. The immunity of the individual is compromised because they have fewer white blood cells in their body to fight disease. A similar thing happens in leukaemia and in those having radiotherapy and other diseases such as viral hepatitis and multiple myeloma.

There are two main types of HIV. These are:

- HIV-1: the most common type found worldwide. There are over 60 different epidemic strains of HIV-1 in the world. But usually different regions are dominated by only one or two of them[101].
- HIV-2: this is found mainly in Western Africa, with some cases in India and Europe[102].

Both of these groups break down into a multitude of further subtypes which go way beyond this writer's Winnie-the-

---

[100] Beatrice Hahn, MD.
www.med.upenn.edu/apps/faculty/index.php/g275/p8418831

[101] www.medicalnewstoday.com/articles/269510.php

[102] www.avert.org/about-hiv-aids/what-hiv-aids

Pooh sized brain.

Causes, Signs and Symptoms

www.avert.org describes three stages to the disease.

1. Acute primary infection – This produces the first set of symptoms described below.
2. The asymptomatic stage – Once the first stage, the **seroconversion** stage, is over, people begin to feel better. HIV might well not display itself again for many years, between 2 and 15 years in some cases[103] [104].
3. Symptomatic HIV infection – 'During the third stage of HIV infection there is usually a lot of damage to your immune system. At this point, you are more likely to get serious infections or bacterial and fungal diseases that you otherwise would be able to fight off. These infections are referred to as 'opportunistic infections'. If a person is experiencing opportunistic infections they are now said to have AIDS[105]'.

According to www.aids.gov, the signs and symptoms for infection are:

- Fever
- Chills
- Rash
- Night sweats
- Muscle aches
- Sore throat

---

103 HIV/AIDS, Poverty and Pastoral Care and Counselling By Vhumani Magezi

104 http://www.who.int/mediacentre/factsheets/fs360/en/

105 www.avert.org/about-hiv-aids/symptoms-stages

- •Fatigue
- •Swollen lymph nodes
- •Mouth ulcers
- •Headache
- •Upset stomach
- •Joint aches and pains

Some people may experience a flu-like illness within 2-4 weeks after HIV infection. But some people may not feel sick during this stage.

The progression to AIDS according to the same website is shown by:

- •Rapid weight loss
- •Recurring fever or profuse night sweats
- •Extreme and unexplained tiredness
- •Prolonged swelling of the lymph glands in the armpits, groin, or neck
- •Diarrhoea that lasts for more than a week
- •Sores of the mouth, anus, or genitals
- •Pneumonia
- •Red, brown, pink, or purplish blotches on or under the skin or inside the mouth, nose, or eyelids
- •Memory loss, depression, and other neurologic disorders.

At this stage also, opportunistic infections begin to emerge. These are such things as pneumonia, meningitis, diarrhoea and encephalitis, due to the 'liberation' of bacteria and viruses by an immunocompromised system.

This is all a bit vague. Night sweats can be attributed to thyroid problems or the menopause, rashes to any number of causes from a change in fabric softener to the painkillers you took that morning. Fatigue could be down to diabetes or burning the candle at both ends. Mouth ulcers are a sign of being run down or eating your crisps too quickly. It's not until you actually start reaching the later stages of HIV, AIDS, that the symptoms become more aggressive, more specific, but once again, it is all a bit vague.

My point is, these have to be observed in combination with lifestyle. As with any sexually transmitted disease, any Thursday night in town can end up with crabs or chlamydia (for the literalists out there, no it doesn't actually *have* to be a Thursday night). The individual has to take the same precautions that they would with any STD.

I don't however want to lump this in with the sexual diseases, because that gives a false impression of the disease and those with it. I have nursed people who caught HIV through blood transfusion. I'm not trying to indicate culpability, nobody deserves this; I'm trying to indicate vulnerability.

HIV is found in the following body fluids of an infected person: semen, blood, vaginal and anal fluids and breast milk. It **cannot be transmitted** through sweat, saliva or urine.

Neither can it be caught from toilets, holding hands, kissing, sharing cutlery, insect bites or from someone with HIV who is in the same country as you that you don't meet, talk to on the phone or who flies over you in an aeroplane. The myths that arose out of those fears of the 80s were ridiculous and yet believed by millions.

Treatment and Prevention

The simple truth is, don't make yourself vulnerable to the disease. The greatest ally that we have is education. It is however, as usual, the poor and uneducated that get left behind, which more often than not translates into those same countries as end up on all the lists: the 'third-world' countries.

> 'Behaviours and conditions that put individuals at greater risk of contracting HIV include:
>
> - having unprotected anal or vaginal sex;
> - having another sexually transmitted infection such as syphilis, herpes, chlamydia, gonorrhoea, and bacterial vaginosis;

- sharing contaminated needles, syringes and other injecting equipment and drug solutions when injecting drugs;
- receiving unsafe injections, blood transfusions, medical procedures that involve unsterile cutting or piercing; and
- experiencing accidental needle stick injuries, including among health workers.[106]'

With early diagnosis, by a simple blood test, and effective antiretroviral treatment, people with HIV can live a normal, healthy life.

- 'Antiretroviral therapy (ART) is the use of HIV medicines to treat HIV infection. People on ART take a combination of HIV medicines (called an HIV regimen) every day.
- ART is recommended for everyone infected with HIV. People infected with HIV should start ART as soon as possible. ART can't cure HIV, but HIV medicines help people infected with HIV live longer, healthier lives. ART also reduces the risk of HIV transmission.
- Potential risks of ART include unwanted side effects from HIV medicines and drug interactions between HIV medicines or between HIV medicines and other medicines a person is taking. Poor adherence—not taking HIV medicines every day and exactly as prescribed—can lead to drug resistance and treatment

[106] www.who.int/mediacentre/factsheets/fs360/en/

failure.[107]'

The World Health Organisation breaks treatment down into three areas:

1. ART (Antiretroviral Therapy) as prevention – The risk of transmitting the disease to a partner can be reduced by up to 96% by using an effective ART programme.

2. Pre-Exposure Prophylaxis (PrEP) for HIV-negative partner – This is where antiretroviral drugs are used by people without HIV to prevent their acquisition of the disease.

3. Post-Exposure Prophylaxis for HIV (PEP) – this is the use of the ARV drugs within 72 hours of exposure to HIV, along with counselling and testing. I'll be honest, I wouldn't fancy trusting my life to this.

By end-2015, 17 million people living with HIV were receiving ART globally which meant a global coverage of 46% (43–50%)[108].

The Social Implications of HIV/AIDS

HIV mutates and survives. It is a very adaptable disease. It will not go away any time soon. Over 35 million people have died from the disease so far (contrary to what the media would have us believe, not just the rich and famous).

In 2015 alone 1.1 million people died from AIDS related illness. There were approximately 36.7 million people living with

---

107 www.aidsinfo.nih.gov/education-materials/fact-sheets/21/51/hiv-treatment--the-basics

108 www.who.int/mediacentre/factsheets/fs360/en/

HIV at the end of 2015 with an average 2.1 million people becoming newly infected with HIV in 2015 globally. It is estimated that currently only 54% of people with HIV know their status. Between 2000 and 2015, new HIV infections have fallen by 35%, AIDS-related deaths have fallen by 28%[109].

When the disease came to our attention in 1981, its effect upon the homosexual community was the most highlighted aspect of it. It made it more newsworthy[110].

This newsworthiness armed the bigots. Paul Cameron, an American *psychologist* used the AIDS crisis to suggest that "the extermination of homosexuals" might become necessary. The next year he was the co-author of a pamphlet which blamed gay men for the epidemic and called for a national crackdown on homosexuals. Paul Buchanan of Ronald Regan's (and he was a *President* of the good ole freedom-lovin' US of A) staff, called AIDS, 'nature's revenge on gay men'[111]. He was the Communications Director at the Nut House…sorry…White House, the implication being that his statement was first vetted and approved by the White House.

Violence against the gay communities also increased, no doubt spurred on by the bravado lent to them by the morons in the White House.

'VIOLENCE AGAINST HOMOSEXUALS RISING, GROUPS SEEKING WIDER PROTECTION SAY

By WILLIAM R. GREER

Published: November 23, 1986

Attacks on homosexuals appear to have increased

[109] www.who.int/mediacentre/factsheets/fs360/en/

[110] Image: www.gayinthe80s.com/2013/01/1984-85-media-aids-and-the-british-press/aids-press-collage/

[111] www.splcenter.org/fighting-hate/intelligence-report/2005/history-anti-gay-movement-1977

> sharply around the nation in the last three years as homosexuals have become more vocal in their pursuit of civil rights and more visible because of publicity surrounding the spread of AIDS.
>
> Law-enforcement agencies do not record crimes against homosexuals as a specific category. In several cities, however, homosexuals have formed organizations to document what they say is growing violence against them, to lobby for more protection and to counsel the victims.[112]'

It's not only the impact upon the gay community that should be noted. Hospitals and nursing homes, undertakers and mortuaries, in fact any health-centred organisation, had to change their working practices. New health and safety laws were introduced, hate laws introduced because of the rank stupidity of some sections of society. The perception by the public of healthcare changed. Could they still safely have blood transfusions? Could they go into hospital and come out without AIDS (don't worry, MRSA has that covered)?

This report, from the *Nigerian Medical Journal* of October-December 2011, highlighted the pervasiveness, both psychological and physical, of the disease:

> 'For prolonged duration and severity of disease, higher proportion of indoor patients reported loss of job, decreased family income, increased expenditure for care seeking, and faced greater economic consequences, reflected by selling assets. Loss of job was mainly due to illness (86.8%), disclosure of sero-status (13.2%), and predominantly among skilled workers. Assets were sold mainly to meet the cost of own illness for

112 www.nytimes.com/1986/11/23/us/violence-against-homosexuals-rising-groups-seeking-wider-protection-say.html?pagewanted=all

> indoor patients, but more to meet the expenditure for husband's illness…High school dropout seen in both groups was mainly due to economic reasons. HIV/AIDS status was known to other members of family for 84.8% of indoor patients out of which 15.4% experienced rejection by family members. Out of 72 ever (sic) married women indoor patients whose in-laws were aware of their HIV/AIDS status, 41.7%, 40.9%, and 33.33% reportedly were blamed for spouse's illness, and had strained relation with in-laws and spouse, respectively.[113]'

Much like Ebola, people are still joking about AIDS and HIV. The novelty, class differences, educational frailties, inbuilt bigotries and fear haven't worn off. Maybe that's a good thing. At least fear and bigotry is a form of awareness, which is better than nothing, unless these people also think that they are immune.

Why have I chosen these four diseases? Two of them, by most people, are thought to be dead or, at the very least, incredibly distant and irrelevant – plague and cholera. The other two - HIV and Ebola - have changed our lives, the way we live. They have woken us up to the consequences of our actions and made us realise that, far from being a place of safety, the modern world might actually increase our risk.

Bacteria and Viruses mutate. We have only had antibiotics since the 1940s and yet they have increased the speed of disease mutation by a hundred percent, a thousand percent, by almost incalculable amounts. There is, in the back of my mind, the story of Frankenstein, the foreboding that we have become Dr Frankenstein and, by trying to triumph over death, we have created a greater monster that brings a new death of its own. We don't yet fully know the consequences of those well-intentioned actions. Perversely, for all the wonders of antibiotics, they might have actually left us even more vulnerable.

---

[113] www.ncbi.nlm.nih.gov/pmc/articles/PMC3329095/

PART 3

Suicide

The 'S' Word

'...neither ancient Greek nor Latin had a single word that aptly translates our 'suicide,' even though most of the ancient city-states criminalized self-killing[114].'

'The first half of the word is from Latin sui (of oneself).

The second half of the term is from Latin caedre (to cut down, strike mortally, kill).'

www.edenics.net

'Mid-17th century: from modern Latin *suicidium* act of suicide, *suicida* - person who commits suicide, from Latin *sui* of oneself + *caedere* kill.'

www.en.oxforddictionaries.com

'Suicide (n.) "deliberate killing of oneself," 1650s, from Modern Latin suicidium "suicide," from Latin sui "of oneself" (genitive of se "self"). Probably an English coinage...The meaning "person who kills himself deliberately" is from 1728. In Anglo-Latin, the term for "one who commits suicide" was felo-de-se, literally "one guilty concerning himself."'

www.etymonline.com

Why the fuss over the meaning of one word? Over the origins of one word? Over when it was or wasn't first used? Because, I know that, even as you saw the word, it brought up emotions within you, flared opinions, made you struggle to maintain objectivity by pushing aside the heavy weight of

---

114 Attributed to John M Cooper - author and clever chap.

subjectivity, made you curl that inner lip in disdain or mentally reach out to the ghosts of all those who have fallen to this particular sword. We all have an opinion.

The word did not exist in ancient Greece and yet it was regularly, almost openly, practiced and it was much-debated. You will have seen from the definitions at the beginning of the chapter, that in English it is a mid-seventeenth century word, an amalgam, a portmanteau, of Latin words, forced together like two unwilling pandas to create another.

But there is more to it than this. There is more to it than simple etymology because, more than any other form of death, suicide is the one that triggers the righteous and the moral reactions and feeds upon our fears – many, many fears. It infringes upon the sacred, upon the inner conflict that exists within the majority of us between religion, duty and freedom.

The Philosophy of Suicide.

You cannot discuss suicide as a means of death without touching upon the philosophy of the act. If I tried to omit it, it wouldn't just be the elephant in the room, it would be the rare white elephant, the one knitting and tutting, slowly swishing its trunk in the way an angry cat twitches its tail. When we left the room, we would know that there were words left unsaid, hot ashes left to glow, rivers uncrossed. There would be an incompleteness, a vacancy left unfilled. This is because, if we are to discuss motives, which we will, then it touches upon the philosophy, the thoughts and needs, of the individual and the pressure upon that individual to stay alive for the sake of conformity or to die for the need to end.

I will start with one of my favourites, Albert Camus, from *The Myth of Sisyphus*:

> 'There is but one truly serious philosophical problem and that is suicide…Suicide has never been dealt with except as a social phenomenon. On

> the contrary, we are concerned here, at the outset, with the relationship between individual thought and suicide. An act like this is prepared within the silence of the heart, as is a great work of art. The man himself is ignorant of it. One evening he pulls the trigger or jumps…Beginning to think is beginning to be undermined. Society has but little connection with such beginnings. The worm is in man's heart. That is where it must be sought. One must follow and understand this fatal game that leads from lucidity in the face of experience to flight from light…In a sense, and as in melodrama, killing yourself amounts to confessing. It is confessing that life is too much for you or that you do not understand it. Let's not go too far in such analogies, however, but rather return to everyday words. It is merely confessing that that 'is not worth the trouble'…Dying voluntarily implies that you have recognized, even instinctively, the ridiculous character of that habit, the absence of any profound reason for living, the insane character of that daily agitation and the uselessness of suffering.'

Camus attacks life as absurd. No matter what purpose you attach to it, it is, ultimately, meaningless. What was here today is gone tomorrow. He also suggests that the individual has little control over the act, that he is forced into it unconsciously. It is an individual decision in which society takes no part except perhaps as a final stimulus towards the events. Once that worm of decay is inside you, however, there is an inevitability to your end. Merely thinking about life, realising that it is meaningless or insurmountable in its suffering, undermines the will and the reason to live. There is no logical motive for staying alive.

Camus is not alone in questioning the validity of life, of enduring pointless suffering, of being open to the admission that life is too much to endure and that it must be ended.

Plato, who was anti-suicide and described the act as

cowardly, disgraceful and lazy, said that there were four circumstances under which suicide was acceptable.

> (1) when one's mind is morally corrupted and one's character can therefore not be salvaged (Laws IX 854a3–5),
> (2) when the self-killing is done by judicial order, as in the case of Socrates,
> (3) when the self-killing is compelled by extreme and unavoidable personal misfortune, and
> (4) when the self-killing results from shame at having participated in grossly unjust actions (Laws IX 873c-d). Suicide under these circumstances can be excused, but, according to Plato, it is otherwise an act of cowardice or laziness undertaken by individuals too delicate to manage life's vicissitudes[115].'

He comes close to Camus (or Camus comes close to him), but pulls out at the last moment, skirts around the idea that life, quite simply, might be too much. The worm inside your heart is not enough for Plato. A good backbone and a stiff upper lip were what were needed in the absence of (1) to (4).

Thomas Aquinas, he of the sainthood, in *Summa Theologica,* argued that suicide was not acceptable because it violated 'three duties: one's duty to God, to others and to oneself.'

> 'Now every man is part of the community, and so, as such, he belongs to the community. Hence by killing himself he injures the community, as the Philosopher declares (Ethic. v, 11).'

---

[115] http://plato.stanford.edu/entries/suicide/ Cholbi, Michael, "Suicide", The Stanford Encyclopedia of Philosophy (Summer 2016 Edition), Edward N. Zalta (ed.)

This is echoed a few centuries later by John Donne (1624)[116]:

> 'No man is an island,
> Entire of itself,
> Every man is a piece of the continent,
> A part of the main.
> If a clod be washed away by the sea,
> Europe is the less.
> As well as if a promontory were.
> As well as if a manor of thy friend's
> Or of thine own were:
> Any man's death diminishes me,
> Because I am involved in mankind,
> And therefore never send to know for whom the bell tolls;
> It tolls for thee.'

It elevates the concept that man and community are inseparable, that what happens to a person happens to those around him. The ripples of consequence will diminish the further they travel, but those closest will suffer.

Ironically, one of the world's most famous suicides, Ernest Hemingway, used this poem in his book *For Whom the Bell Tolls*, his novel about the Spanish Civil War.

It wasn't until 1755, when David Hume in *Of Suicide* began to put forward a defence of suicide, that the world began to take on an alternative perspective.

> 'According to the Thomistic [Thomas Aquinas] argument, suicide violates the order God established for the world and usurps God's prerogative in determining when we shall die. Hume's argument against this thesis is intricate and rests on the following considerations:

[116] Devotions upon Emergent Occasions - Meditation XVII

1. If by the 'divine order' is meant the causal laws created by God, then it would always be wrong to contravene these laws for the sake of our own happiness. But clearly it is not wrong, since God frequently permits us to contravene these laws, for he does not expect us not to respond to disease or other calamities. Therefore, there is not apparent justification, as Hume put it, for God's permitting us to disturb nature in some circumstances but not in others. Just as God permits us to divert rivers for irrigation, so too ought he permit us to divert blood from our veins.
2. If by 'divine order' is meant the natural laws God has willed for us, which are (a) discerned by reason, (b) such that adherence to them will produce our happiness, then why should not suicide conform to such laws when it appears rational to us that the balance of our happiness is best served by dying?
3. Finally if by 'divine order' is meant simply that which occurs according to God's consent, then God appears to consent to all our actions (since an omnipotent God can presumably intervene in our acts at any point) and no distinction exists between those of our actions to which God consents and those to which He does not. If God has placed us upon the Earth like a "sentinel," then our choosing to leave this post and take our lives occurs as much with his cooperation as with any other act we perform.

Furthermore, suicide does not necessarily violate any duties toward other people, according to Hume. Reciprocity may require that we benefit society in exchange for the benefits it provides, but surely such reciprocity reaches its limit when by living we provide only a "frivolous advantage" to society at the expense of significant harm or suffering for ourselves. In more extreme situations,

> we are actually burdens to others, in which case our deaths are not only "innocent, but laudable."
> Finally, Hume rejects the thesis that suicide violates our duties to self. Sickness, old age, and other misfortunes can make life sufficiently miserable that continued existence is worse than death. As to worries that people are likely to attempt to take their lives capriciously, Hume replies that our natural fear of death ensures that only after careful deliberation and assessment of our future prospects will we have the courage and clarity of mind to kill ourselves. In the end, Hume concludes that suicide "may be free of imputation of guilt and blame."[117]'

'Why should I prolong a miserable existence, because of some frivolous advantage which the public may perhaps receive from me?' he said in his *Essays on Suicide and the Immortality of the Soul.*

I think I would have liked Mr Hume.

Nowadays, attitudes generally have changed. We don't refuse to bury the suicide on consecrated ground, we don't put stakes through their hearts or bury them on crossroads, as superstition and religion once dictated that we should.

> 'In the Middle Ages, suicide was often regarded as the result of diabolical temptation, induced by despair or madness. Savage penalties were inflicted on the dead body - such as dragging it through the streets where the deceased had lived, and hanging it. The estates of these persons were confiscated, and Christian burial was forbidden. Sometimes, the corpse of a suicide was buried at a busy crossroads (in order to confuse the spirit), pinned down by a

---

[117] http://plato.stanford.edu/entries/suicide/ Cholbi, Michael, "Suicide", The Stanford Encyclopedia of Philosophy (Summer 2016 Edition), Edward N. Zalta (ed.)

> wooden stake through the chest - thus preventing, it was hoped, the spirit emerging to bother the living.[118]'

It remains divisive however, especially when we talk about assisted suicide. The arguments divide clearly into two camps and, as with many moral arguments, the two tribes are fairly immovable on the subject. The interesting thing is that the arguments remain essentially the same as when Plato, Hume and Aquinas were pronouncing to the world.

I have had many discussions about assisted suicide and remain one of the namby-pamby few who sit on the fence. I am influenced by my perspective as a nurse, but not for the reasons that you might think. I don't see life as sacred, as God-given, as not owned by the liver. As a nurse I was employed to preserve life or see it to a comfortable and natural (as possible) end. It was to minimise the physical and psychological suffering of the individual until the final resolution, whatever that resolution might have been – yes, some patients did actually survive my care. Partly on this basis, I was in favour of 'euthanasia' as a way to end or prevent suffering.

But I also saw the greed of families and know that, given the chance, they would have used the euthanasia route as a way to Mum's house or her jewellery or Dad's money. The worst of human nature would come to the surface and I feel sure that, under the right circumstances, 'euthanasia' would simply be legalised murder. It's all very well to say that contingencies will be put in place to avoid this eventuality, but there are ways around contingencies and the rights of the dying individual are easily cast aside. I admit to feeling torn on this issue.

It is an age-old debate that will rage as long as humans thrive. I'm glad about that because I think that if we fail to question our existence, our place in the world, then purpose dissolves. I love Camus and his work, but in this case, I would slightly disagree with him; the worm in my heart propels me, but

---

[118] http://soars.org.uk/index.php/about/2014-06-06-18-57-53

only just.

Interestingly, we still see an attempt to commit suicide as an indication of some sort of mental illness. It doesn't occur to us that it might actually be a rational, well thought-out, happily-made decision.

Out of interest, today (17 September 2016) the BBC put out a story about the first Belgian minor to be granted euthanasia. The 17-year-old was said to have ben 'suffering unbearable pain'.

> 'Belgium is the only country that allows minors of any age to choose euthanasia. They must have rational decision-making capacity and be in the final stages of a terminal illness. The parents of the under-18-year olds must also give their consent. Euthanasia commission head Wim Distelmans said the teenager was "nearly 18". He said doctors used "palliative sedation", which involves bringing patients into an induced coma, as part of the process.'

Usefully, the article also goes on to differentiate between the three types of 'euthanasia:

1. Euthanasia is an intervention undertaken with the intention of ending a life to relieve suffering, for example a lethal injection administered by a doctor
2. Assisted suicide is any act that intentionally helps another person kill themselves, for example by providing them with the means to do so, most commonly by prescribing a lethal medication
3. Assisted dying is usually used in the US and the UK to mean assisted suicide for the terminally ill only, as for example in the Assisted Dying Bills recently debated in the UK

They are important distinctions and I suspect will be the basis for much future legal wrangling. The article further said:

- The Netherlands, Belgium and Luxembourg permit euthanasia and assisted suicide
- Switzerland permits assisted suicide if the person assisting acts unselfishly
- Colombia permits euthanasia
- California last year joined the US states of Oregon, Washington, Vermont and Montana in permitting assisted dying
- Canada passed laws allowing doctor-assisted dying in June of this year

A Definition of Suicide

Suicide is the wilful and voluntary act of a person who understands the physical nature of the act, and intends by it to accomplish the result of self-destruction. Suicide is the deliberate termination of one's existence, while in the possession and enjoyment of his [their] mental faculties. Self-killing by an insane person is not suicide.

www.thelawdictionary.org/suicide

'The National Statistics definition of suicide includes deaths given an underlying cause of intentional self-harm or an injury/poisoning of undetermined intent. However, this cannot be applied to children due to the possibility that these deaths were caused by unverifiable accidents, neglect or abuse. Therefore, only persons aged 15 years and over are included in the suicide figures.'

ONS Suicides in the United Kingdom, 2013 Registrations - 19 February 2015

UK definition of suicide

- Intentional self-harm
- Injury/poisoning of undetermined intent
- Sequelae of intentional self-harm/injury/poisoning of undetermined intent

The above definition is from www.samaritans.org. It is quite difficult to find a definitive definition (he said with a sense of irony) but, out of those looked at, the Samaritans seem to talk the most sense. They do however take their definitions from the Office for National Statistics. The Samaritans tend to magically turn statements from gobbledegook into English.

Gimme Some Stats, Stat…

- In 2013, 6,233 people in the UK committed suicide. This was of people aged over fifteen.
- This works out as 11.9 deaths per 100,000 people.
- The highest suicide rate was among men aged 45-59.
- Regionally, the highest suicide rate was in the North-East of England at 13.2 per 100,000. The lowest was in Yorkshire and the Humber with 9.7 per 100,000, beaten only by the ecstatically happy Welsh with 9.2 per 100,000. (see regional map)

- Of the 6,233 deaths in England and Wales in 2013, 4,858 were male. 1,375 were female.
  The highest age group for female suicides was 45-59.[119]
- This breaks down as 78% male and 22% female.
- It has always been the case that more men commit suicide than women.

- 'Overall, there were 6,122 suicides in the UK in 2014, 120 fewer than in 2013 – a 2% decrease. The male suicide rate was more than three times higher than the female rate, with 16.8 male deaths per 100,000 compared with 5.2 female deaths.[121]'
-
- In 2014 there were 6,122 suicides of people **aged 10** and over**. Repeat: Aged 10** and over. When did life in this country become so shite for ten-year olds that they had to kill themselves?

SUICIDE RATE BY REGION PER 100,000	
WALES	15.6
NORTH EAST	13.8
SOUTH WEST	12.5
NORTH WEST	12.3
YORKSHIRE AND THE HUMBER	11.6
SOUTHEAST	11.4
ENGLAND	10.7
WEST MIDLANDS	10.4
EAST MIDLANDS	10.2

[119]Stats: www.independent.co.uk, Thursday 19 February 2015

EAST OF ENGLAND	9.4
LONDON	7.9

> '**98 CHILDREN AGED 10-14 HAVE KILLED THEMSELVES IN LAST DECADE**... The records show 98 children under 15 killed themselves in the UK from 2005 to 2014, 59 boys and 39 girls... experts had known for some time that depression started for some at an early age, and there were a "plethora of reasons" why children took their own lives. He said children felt "trapped or ashamed" by whatever was driving their suicidal thoughts.[120]'

- The highest suicide rate in the UK in 2014 was among men aged 45 to 59.
- The most common suicide method in the UK in 2014 was hanging, which accounted for 55% of male suicides and 42% of female suicides.
- Men aged 45 to 59 had the highest suicide rate in 2014 for the second year in a row with a rate of 23.9 deaths per 100,000 population.
- Men aged 30 to 44 had the second highest suicide rate, at 21.3 deaths per 100,000 population in 2014.
- In 2014, the age group with the highest suicide rate for females was 45- to 59-year-olds. This has been constant since 2003[121].

---

[120] www.theguardian.com. Mark Tran. 4 February 2016

[121] Office for national statistics Suicides in the United Kingdom, 2013 Registrations and Samaritans.org

- •Divorced men commit suicide at a rate 10 times greater than divorced women.
- Teen boys commit suicide at a rate 4 times greater than girls.
- There has been a 128% increase in suicides among children aged 10-14 since 1980 (in the USA)[122].
-

According to the ever-reliable Office for National Statistics:

-
- In 2017 there were 5,821 suicides registered in the UK, an age-standardised rate of 10.1 deaths per 100,000 population.
- The UK male suicide rate of 15.5 deaths per 100,000 was the lowest since our time-series began in 1981; for females, the UK rate was 4.9 deaths per 100,000, this remains consistent with the rates seen in the last 10 years.
- Males accounted for three-quarters of suicides registered in 2017 (4,382 deaths), which has been the case since the mid-1990s.
- The highest age-specific suicide rate was 24.8 deaths per 100,000 among males aged 45 to 49 years; for females, the age group with the highest rate was 50 to 54 years, at 6.8 deaths per 100,000.
- Scotland had the highest suicide rate in Great Britain with 13.9 deaths per 100,000 persons, and England the lowest with 9.2 deaths per 100,000.

[122] www.psychologytoday.com

These are fairly staggering statistics open to, no doubt, wide interpretation. Hey can be (mis)interpreted until they are inside out, but they do give an insight into society at a specific time. For example, why do so any men between 45 and 49 years kill themselves? Is it high rates of unemployment and the likelihood that they will not find a job again because of their age? Is it divorce? Women tend to have closer and larger social circles than men; perhaps this indicates a better support system available to them in times of need. Do these men feel that they have not met up, in some way with the expectations of their society and therefore see no other way out because of the stigma attached to their perceived failures? However, male suicides were down on previous years, so does his mean that there is a better support system out here for them, that they are willing to accept therapy before they attempt to do away with themselves? Or are they surviving through crisis intervention, post-attempt? Either way, there is an implication that more support exists.

Reasons for Suicide

1) They're depressed

There is without doubt a link between depression and suicide. If the associations with suicide risk factors are examined, they are also closely linked with depression, which in turn are linked with social factors.

- drug and alcohol misuse
- history of trauma or abuse
- unemployment
- social isolation
- poverty
- poor social conditions
- imprisonment
- violence

- family breakdown[123].

The BBC on 12 June 2014 reported that

> 'the economic crisis in Europe and North America led to more than 10,000 extra suicides, according to figures from UK researchers. A study, published in the British Journal of Psychiatry, showed "suicides have risen markedly". The research group said some deaths may have been avoidable as some countries showed no increase in suicide rate. Campaign groups said the findings showed how important good mental health services were. The study by the University of Oxford and the London School of Hygiene & Tropical Medicine analysed data from 24 EU countries, the US and Canada. It said suicides had been declining in Europe until 2007. By 2009 there was a 6.5% increase, a level that was sustained until 2011. It was the equivalent of 7,950 more suicides than would have been expected if previous trends continued, the research group said.'

On July 8 2011, nhs.uk reported, "'Suicide rates have risen sharply across Europe since the banking crisis'". The Independent today reported that, 'a study has found that Britain has been affected particularly hard, experiencing an 8% rise in suicide rates between 2007 and 2009. Ireland and Greece, two of the countries reported to be in greater financial difficulty, have seen suicides rise 13% and 16% respectively.'

More than 90% of suicides are associated with a psychiatric disorder. The highest rates of suicide were associated with depressive disorders[121]. In the USA, over 50% of all people who die by suicide suffer from major depression. If one includes alcoholics who are depressed, this figure rises to over 75%[124].

---

[123] www.mentalhealth.org.uk/a-to-z/s/suicide

[124] http://theovernight.donordrive.com/?fuseaction=cms.page&id=1034

2) They're Psychotic.

This is a rather 'direct' way of describing particular types of mental health problems, but it's accurate. Psychosis is defined as a mental illness in which false beliefs are held and commonly accompanied by illusions and/or delusions and/or hallucinations of either/and an aural or visual type. It is often simply defined as 'a loss of touch with reality'.

> 'Suicide is a major cause of death among people who develop psychotic illness…A recent meta-analysis found that 18.4% of first-episode patients had self-harmed or attempted suicide prior to initial treatment, while the proportion of patients who committed acts of self-harm in the period after initial treatment was 11.4%[125].'

In *Suicidal Behaviour In Patients With Schizophrenia And Other Psychotic Disorders*, the authors found that patients with schizophrenia and other psychotic disorders were at greater risk of suicide attempts than those without. In a group of 1048 patients, they found that 30.2% had attempted suicide at some stage in their life, with 7.22% admitting that they had attempted suicide within one month prior to admission. The lifetime risk of suicide for people with schizophrenia and bipolar disorder has been estimated at 10-15%. Among patients with psychosis, the lifetime risk of self-harm is estimated to be as high as 20-50%[126].

Suicide and mental disorders: distribution of diagnoses in studies with general population[127]

---

[125] High rates of suicide attempt in early-onset psychosis are associated with depression, anxiety and previous self-harm. Olav Nielssen, Matthew Large

[126] www.kcl.ac.uk

[127] *Suicide and psychiatric diagnosis: a worldwide perspective.* José Manoel Bertolote1 and Alexandra Fleischmann. World Psychiatry. Oct 2002.

Diagnosis	Percentage of Total
Mood disorders	35.8
Substance-related Disorders	22.4
Personality disorders	11.6
Schizophrenia	10.6
Anxiety/somatoform disorders	6.1
Other DSM axis 1 Diagnosis	5.1
Adjustment disorders	3.6
No Diagnosis	3.2
Organic Mental Disorders	1.0
Other Psychotic Disorders	0.3

It might help to qualify some of these elusive terms. Professionals like to use big words in order to maintain their exclusivity. Let's break that wall down.

- Mood Disorders – any disorder that affects the mood, such as depression or bipolar disease.
- Substance-Related Disorders – There have been links to mental illness and cannabis use, to alcohol and depression. This usually refers to addictive substances that cause unacceptable or self-harming behaviour.
- Personality Disorders – nicely vague this one. I'll use the nhs.uk definition: 'Personality disorders are conditions in which an individual differs significantly from an average person, in terms of how they think, perceive, feel or relate to others.' This possibly raises more questions than it answers.
- Schizophrenia – 'Schizophrenia is a severe brain disorder in which people interpret reality abnormally. Schizophrenia may result in some combination of hallucinations, delusions, and extremely disordered thinking and behaviour.[128]'

---

[128] www.mayoclinic.org -

- Anxiety/Somatoform Disorders (somatic symptom disorder) – A somatoform disorder is a mental illness that is presented with bodily symptoms. 'The patient must also be excessively worried about their symptoms, and this worry must be judged to be out of proportion to the severity of the physical complaints themselves. A diagnosis of somatic symptom disorder requires that the subject have recurring somatic complaints for at least six months.[129]' There is no fakery in this illness. The individual has the pain, but there is no discernible physical cause. Anxiety – this is now an enormous burden upon younger people. Quite why it should be so prevalent today as compared to say, forty years ago, could be down to such things as better diagnosis, a greater willingness in society to disclose personal information, greater stress placed upon the individual by modern society or more widespread drug use. It is essentially an extreme sense of apprehension which can present itself in physical form such as palpitations or sweating or might even produce panic attacks.
- Other DSM Axis 1 Diagnosis – DSM stands for Diagnostic and Statistical Manual of Mental Disorders. It is simply a way of classifying mental illness and making the compiler feel special. 'This is the top-level of the DSM multiaxial system of diagnosis. It represents acute symptoms that need treatment; Axis I diagnoses are the most familiar and widely recognized (e.g., major depressive episode, schizophrenic episode, panic attack).[130]'
- Adjustment Disorders – These involve things such as stress or the physical symptoms that occur after going through a stressful life event eg, the death of a loved one.
- Organic Mental Disorders – These are mental illnesses

---

129 Oyama O, Paltoo C, Greengold J (November 2007). *"Somatoform disorders". American Family Physician.*

130 www.psyweb.com

which are due to a physical or medical cause rather than a psychiatric cause. This can be anything from concussion to stroke to carbon monoxide poisoning. It does not, however, include dementia.

- Other Psychotic Disorders – exclusive of Schizophrenia; a mental illness in which false beliefs are held and commonly accompanied by illusions and/or delusions and/or hallucinations of either/and an aural or visual type. It is often simply defined as 'a loss of touch with reality'. There are quite a few, too numerous for '*A Beginner's Guide…*', but easily searchable on the net should you wish to pursue the subject.

3) They're impulsive

Aren't we all sometimes? This is often related to drugs and alcohol, those things that break down the inhibitions and bring regret etc to the fore. Most people who attempt suicide for this reason are often lucky enough to regret it. If they're not lucky, they go into category 6). The fact that it was done on impulse and regret/remorse/shame is shown afterwards, does not mean that there is not an underlying problem. They still trod the path.

4) They're crying out for help

It is really as it says. There is often no intent to commit suicide, they don't believe that they will die, but they genuinely feel that they have nowhere to turn, that their situation is excessively burdensome or that no one is listening. The danger is that they too could end up in category 6). People don't realise just how easily paracetamol screws your liver. Just because it's a fifteen or sixteen-year old girl or lad who is crushed by the failure of their first love, does not make the experience any less valid. We all cope with things in different ways, depending upon our backgrounds. The first experience of any failure can be extremely crushing. It should not be invalidated by the world-weariness of the likes of me.

5) They have a philosophical desire to die

Controversial. I feel that people have a right to die if they wish to. One must assume/have a check system to ensure that they understand the consequences of their act, the ripples of their action. We have discussed this at the beginning of the chapter, especially with regard to assisted suicide etc. I don't think that we discuss it enough. The difficulty comes in differentiating the philosophical suicides from the depressives. That is why psychologists and psychiatrists get paid so damned much.

6) They've made a mistake[131]

It happens, I'm afraid. The important thing about this is prevention, making help accessible, making people feel that they are in a safe enough environment to talk and educating people about not creating an environment where people feel the need to create such fatal errors eg bullying. Once again though, the mistake might be indicative of a deeper problem.

7) Perceived Burdensomeness

This is an interesting and very subjective category. The individual, for whatever reason, feels that they are such an excessive burden upon those around them that those people would be better off without them. Studies have shown that those who have attempted suicide or who have left notes behind to explain their actions, do express a high degree of burdensomeness.

8) Low Belonging/Social Alienation

This is very pertinent in today's society. I am thinking of course of the alienation that some ethnic or religious minorities

---

131 www.psychologytoday.com. Alex Lickerman M.D. *The Six Reasons People Attempt Suicide*

must feel in order to drive them to acts of terrorism. We push hard the subjects of equality and diversity in schools and colleges and workplaces, but one has to wonder how much of this is merely lip service to a bad-nanny state. Even in the school arena, within small groups a million miles from terrorism, the same alienation and exclusion occurs which causes individuals to be on the fringes of society and either turn the alienation in on themselves in the form of mental illness or cause them to express it in more dangerous ways to the perceived offenders – you only have to look at the shocking amount of school shootings in Britain and America to see the fallout from this.

9) Acquired Ability to Enact Lethal Self-Injury[132]

This states that 7) and 8) come together to such an extent that it produces in the individual a desire to die. For these two to come together, so the theory goes, it must have the dreadful sounding 'acquired capacity for lethality'. In *The Interpersonal-Psychological Theory of Suicidal Behavior: Current Empirical Status* by Thomas Joiner, PhD, he says that 7) and 8) are 'not sufficient to ensure that desire will lead to a suicide attempt. Indeed, in order for this to occur, the theory suggests a third element must be present: the acquired ability for lethal self-injury.' This is essentially the repetition of the factors that lead to self-harm – bullying, exclusion etc. The twist on this is that this exposure to such stimulus actually makes the individual strong enough to kill themselves; the fear in which they live actually gives them the courage to kill themselves. Their fear takes away their fear. Wowser! Mind-bending and fascinating.

---

[132] www.apa.org/science/about/psa/2009/06/sci-brief.aspx

[133]Methods of Suicide

UK SUICIDE RATE BY REGION PER 100,000	
WALES	15.6
NORTH EAST	13.8
SOUTH WEST	12.5
NORTH WEST	12.3
YORKSHIRE AND THE HUMBER	11.6
SOUTHEAST	11.4
ENGLAND	10.7
WEST MIDLANDS	10.4
EAST MIDLANDS	10.2
EAST OF ENGLAND	9.4
LONDON	7.9

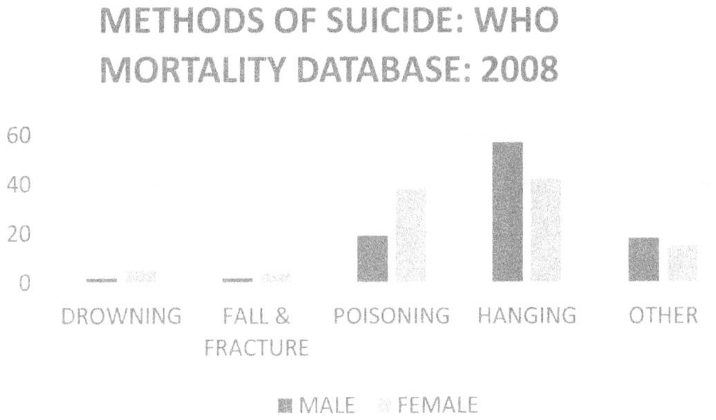

[133] Image: www.ons.gov.uk *Suicides in the United Kingdom: 2014 Registrations.* February 2016

MALES

Country	Year	Other Poisoning	Pesticides	Hanging	Drowning	Firearms	Falls	Other
USA	1999-2002	7.1	0.3	20.4	0.9	60.6	1.9	8.8
UK	2001-2004	14.7	0.4	55.2	2.4	3.5	2.9	20.8

FEMALES

Country	Year	Other Poisoning	Pesticides	Hanging	Drowning	Firearms	Falls	Other
USA	1999-2002	31	0.5	16.9	2.1	35.7	3.4	10.5
UK	2001-2004	41.1	0.3	35.9	4.7	0.6	3.7	13.9

What can we tell from these tables? What are the consistencies or indeed the inconsistencies?

It's clear that the most 'popular' form of suicide is hanging, in the UK that is. In the US, the gun wins, followed by poisoning, which would be such as an overdose of medication.

But what about the important difference between men and women? How does their natural make-up affect the way they decide to end it all? And what about those areas of the world that have the greatest gaps between rich and poor or employment and unemployment? Are those with less money happier or do people think as I do, that I would be *really really* happy with a load of dosh?

This difference is referred to as 'The Gender Paradox in Suicide' by Silvia Canetto, in *Suicide and Life Threatening Behavior.*

There is no doubt of the differences between Venus and Mars. Men choose different ways to kill themselves and they have different motives. They also use suicide more often than women. Men are over 3 times more likely than women to commit suicide.

'Male suicide rate in UK discovered to be 3½ times that of women

Samaritans says men at greatest risk in 40-44

> **age bracket as Office for National Statistics reports 4,590 male suicides in 2012.**[134]
> '...the highest suicide rate was among men aged 40 to 44, at 25.9 deaths per 100,000. This bore out the...studies, which have found middle-aged men of low socioeconomic status to be most at risk. "They will grow up expecting by the time they reach mid-life they'll have a wife who will look after them and a job for life in a male industry," she said. "In reality they may find that they reach middle age in a very different position. Society has this masculine ideal that people are expecting to live up to. Lots of that has to do with being a breadwinner. When men don't live up to that it can be quite devastating for them."'

So, expectations play a large part in the outcome of the individual and this can differ from culture to culture. In the west, we have certain expectations of ourselves and of others – we will be successful at school, then at university, then in our job. Alongside this, we will have a successful marriage, 2.4 children and a shiny BMW on the drive. This puts tremendous pressure upon the individual. The BMW boy is a representation of one strata of society, weighed by class and financial expectations. As you move along the pay and social scales, there will be different expectations educationally, vocationally and matrimonially.

This is the common thread that runs through society – that certain standards or potential or income will be met. If they are not met, then society will question the individual as to why they have not been met. Is the man who remains unmarried at thirty-four gay? Is the woman who is unmarried at that age a lesbian? Is the choice of lifestyle over happiness (or vice-versa) deemed to be wrong within that strata of society? Certain of us

---

134 www.theguardian.com February 18 2014

will be expected to join the civil service and trudge into London for the rest of our working lives because that is what that individual's father did. When those expectations are not met, the resultant pressure is enormous.

> '…of the 5,981 deaths by suicide in the UK in 2012, more than three quarters (4,590) were males. In the US, of the 38,000 people who took their own lives in 2010, 79% were men. It's principally a question of method. Women who attempt suicide tend to use nonviolent means, such as overdosing. Men often use firearms or hanging, which are more likely to result in death.
>
> In the UK, for instance, 58% of male suicides involved hanging, strangulation or suffocation. For females, the figure was 36%. Poisoning (which includes overdoses) was used by 43% of female suicides, compared with 20% of males. A similar pattern has been identified in the US, where 56% of male suicides involved firearms, with poisoning the most common method for females (37.4%).[135]'

And this is a worldwide phenomenon (except in China where women are 40% more likely to commit suicide[136]), not just US and UK based[137].

So why the discrepancy between men and women?

Socioeconomic status has a lot to do with it. Those with lower status are more than ten times more likely to kill

---

[135] www.theguardian.com/science/2015/jan/21/suicide-gender-men-women-mental-health-nick-clegg

[136] Barlow, D. H., and V. M. Durand. Abnormal psychology, an integrative approach. (2011)

[137] Image/Table: www.who.int

themselves[138]. Why? Because the less money you have, the more the small things matter, the less capacity (both intellectually and financially) you have to deal with it. Christmas is not a time of struggle for the wealthy (although it is a myth that more suicides happen at Christmas – it's usually afterwards in January), but for those on low or no wage, it is a great time of stress hammered home by the adverts on TV and the regurgitation of carols and Slade on a twenty-four-hour cycle. When your shortcomings are constantly highlighted, the message does get home.

Unemployment plays into the hands of the gender stereotype as well as the socioeconomic expectations of society. As a man (yes, ladies, even today) I am still expected to have the better job and to earn the greater amount of money. The woman, despite the pivotal role that she now plays within the family structure, is still *perceived* as the secondary partner. Not by me. Don't send me e-mails. This is the gender stereotype soup out of which we have not yet crawled. It is changing, make no bones about that, but unconsciously we still have to get over the innate hump of male superiority and the female as 'support worker'.

Men also seem to take divorce or separation much harder. Once again, we are looking at the individual's inability to live up to the standards expected of them, by themselves and others.

> '…it's been shown that men derive more mental and physical health benefits from marriage than do women (although it's good for both sexes) – so the breakdown of a marriage could lead to more detrimental outcomes for men. That said, there's still a lot of pressure on men to fill out the masculine husband role, whatever socioeconomic class one is in, and the reality is that today this classic role may be somewhat unrealistic. "There is

[138] www.forbes.com/sites/alicegwalton/2012/09/24/the-gender-inequality-of-suicide-why-are-men-at-such-high-risk/#12acdce422f3

> a large and unbridgeable gap between the culturally authorised idea of 'hegemonic masculinity' and the reality of everyday survival for men in crisis," write the authors. One way of taking back one's own masculinity [and control], they suggest, is to take one's own life.[138]'

The other thing that, on the surface, might seem a little inconsequential, is the ability of men to take more pain than women (I admit here and now that my wife is stronger and more dangerous than me and would like to state that this is in no way a challenge). If you look at the comparison between firearms and hanging for both men and women in the previous table, you will notice how high the men's use of these methods is when compared to the women. This is because they are more violent, might involve more pain and involve disfigurement.

Disfigurement plays an important part in the choice of method. It is thought than when men attempt suicide, they mean it, so the method and the message are stamped home in a violent and certain way. They give little thought to how they look. Women care about how they are perceived *post-mortem* and, despite the suicide figures being heavily biased towards men, women actually *think* about suicide more than men. Poisoning is a way to maintain 'looks' and perhaps gives them the opportunity to change their mind. There is a difference in intent when it comes to the thought and the act.

> 'Although suicide rates are lower among women, women lead men two to one in suicide attempts. So, Murphy[139] [George E. Murphy, M.D] says at least 200,000 women are involved in suicide attempts annually. But he points out that attempted

[139] George E. Murphy, M.D. *Why Women Are Less Likely to Commit Suicide. Comprehensive Psychiatry*. From: www.sciencedaily.com November 12 1998

> suicide most often is not an attempt to actually end one's life. Its purpose, he says, is to survive with changed circumstances. "An attempted suicide is not really an attempt at suicide in about 95% of cases. It is a different phenomenon. It's most often an effort to bring someone's attention, dramatically, to a problem that the individual feels needs to be solved. Suicide contains a solution in itself," he says. In attempted suicide, both men and women tend to use methods that allow for second thoughts or rescue. Murphy says that when people intend to survive, they choose a slowly effective, or ineffective, means such as an overdose of sleeping pills. That contrasts to the all-or-nothing means like gunshots or hanging used by actual suicides... women, when they intend to do it, can be just as effective as men in committing suicide. But they aren't so inclined," Murphy says. [140]'

Another important social and psychological aspect is that women are more likely to share; they 'seek feedback and take advice', says Murphy, writing in the journal *Comprehensive Psychiatry*. Men are not. Sharing, in many cases, relieves the sharer of the burden, enough to remove the suicidal intent. Men do not share – we don't even like to go to the doctor. This withholding of stress, this *internalisation* of stress, leads inevitably to damage being inflicted upon the individual, either psychologically or physically. On top of this, the woman takes a much less selfish, more holistic view and considers the impact of her actions upon others.

To be considered in all this is access to the methods of suicide. Where guns are more prevalent eg rural America, there will be more suicides by gunshot. In the UK, we now have a law

---

[140] www.sciencedaily.com/releases/1998/11/981112075159.htm

that stops shops selling more than a certain amount of paracetamol to a single customer – this doesn't of course stop the determined suicide popping from shop to shop in search of death candy. The hope is that those still in doubt will be put off enough by the limitations of sale to think again.

There are also, of course, racial and ethnic disparities; this from www.cdc.gov/violenceprevention:

- Suicide is the eighth leading cause of death among American Indians/Alaska Natives across all ages.
- Among American Indians/Alaska Natives aged 10 to 34 years, suicide is the second leading cause of death.
- The suicide rate among American Indian/Alaska Native adolescents and young adults ages 15 to 34 (19.5 per 100,000) is 1.5 times higher than the national average for that age group (12.9 per 100,000).
- The percentages of adults aged 18 or older having suicidal thoughts in the previous 12 months were 2.9% among blacks, 3.3% among Asians, 3.6% among Hispanics, 4.1% among whites, 4.6% among Native Hawaiians /Other Pacific Islanders, 4.8% among American Indians/Alaska Natives, and 7.9% among adults reporting two or more races.
- Among Hispanic students in grades 9-12, the prevalence of having seriously considered attempting suicide (18.9%), having made a plan about how they would attempt suicide (15.7%), having attempted suicide (11.3%), and having made a suicide attempt that resulted in an injury, poisoning, or overdose that required medical

attention (4.1%) was consistently higher than white and black students.

These are American figures. We in the UK seem a bit coy about displaying our own figures. How odd. Maybe we just don't have that sort of problem in our happy land.

Suicide holds a guilty fascination for us all. We constantly rubberneck the media for such news in the same way that people slow down to see the bloodstain at a road traffic accident.

Our morbid captivation comes from a lack of knowledge, a lack of experience, a lack of understanding, a desperate need to know, a learned and innate preoccupation with death and an unconscious (I hope) superiority in knowing that there is someone weaker than ourselves, somewhere out there, who couldn't hack the system. It makes us feel better about ourselves. It is a guilty pleasure.

It is possibly the most intriguing part of Death and Disease. It is death for the sake of death, to escape life, to enact the philosophical and end the dreadful. It is disease at its most pernicious, invisible and hard-hearted. Unlike plague or cholera or HIV or Ebola, it is self-destructive, voluntary, sought. It contradicts the moral roots and sometimes the laws of society, goes against natural instincts and, in its courage and occasional ingenuity, secretly brings out the admirer in a few of us, though we would never admit it, so constrained are we by the boundaries in which each individual culture, like a strand of DNA, binds itself.

SUICIDES PER 100,000 PER YEAR		
	2009	**2018**
LITHUANIA	28.6	31.9
REPUBLIC OF KOREA	26.3	26.9
BELARUS	24	26.2
KAZAKHSTAN	23.5	22.5
RUSSIA	22.8	31
JAPAN	19.1	18.5

HUNGARY	19	19.1
LATVIA	18.6	21.2
UKRAINE	17.6	22.4
FINLAND	17.1	15.9

Lithuania remains top of the pile 9 years apart – 22.9% of its population lives below the poverty line in 2018; that's a one percent increase on 2017. The Ukraine had undergone several years of Russian oppression. In Japan in 2104, 25,000 people took their own lives.

> This year's report, approved by the Cabinet, shows that the number of people who took their own life declined to 21,897 in 2016, the lowest level in 22 years. But the figures also show that suicide was the top cause of death among people in five age groups from 15 to 39, a trend that stands out amid a decline in other generations.
>
> Tomoko Otake. The Japan Times. May 30 2017.
>
> '"Isolation is the number one precursor for depression and suicide," says Wataru Nishida, a psychologist at Tokyo's Temple University. "When all else fails - some people feel - you can just kill yourself and the insurance will pay out.' Now it's more and more common to read stories about old people dying alone in their apartments," he says. "They are being neglected. Kids used to take care of their parents in old age in Japan, but not any more."'
>
> Wataru Nishida, Temple University

BBC 3 July, 2015

Russia has become notorious for gang-related warfare and its homophobic policies.

> 'Another group recorded over 300 homophobic attacks this year, a more than tenfold rise.
> "We were just having coffee, harming no-one, when men in masks broke-in," Ivan Surok says of one incident, last November. At least one of the attackers was wielding a pellet gun and shot a man in the eye, blinding him; a girl was wounded in her back.
> No-one has yet been prosecuted, part of what Human Rights Watch calls a culture of "widespread impunity". In the cases it documented between 2012 and 2014 only three were brought to court and two led to convictions.
> Since the attack he witnessed, Ivan has carried a pepper spray for protection but no longer feels safe.
> "Homophobes feel like they have a legal basis for their hatred now," Ivan says of the gay propaganda law. "They feel they can beat someone for being gay and they're protected."
> The law - an amendment to child protection legislation - was introduced in several regions before being adopted nationwide'.

BBC, 16 December 2014

There are a thousand reasons why suicide levels remain consistent in these and other countries; it might be social isolation, poverty, political oppression, homophobia, peer pressure, social pressure, culture, mental illness, gender or sexuality, none of them good enough to justify the self-inflicted loss of life.

Humans are the only species since time began that will cause its own extinction. That is a remarkable thought. Whether it be through global warming, war, disease or simply a suicidal impulse born of denial or despair, the truth is, we never really gave ourselves a chance.

PART 4

Accidental Death

'America is an outstandingly dangerous place. Consider this: every year in New Hampshire a dozen or more people are killed crashing their cars into moose. Now correct me if I am wrong, but this is not something that is likely to happen to you on the way home from Sainsbury'.'

(The Great) Bill Bryson - *Notes From a Big Country* (1998)

'Niles, you know as well as I do there are no accidents!'

Frasier Crane - *Bla-Z-Boy* (2001)

Apparently, there are no accidents anymore, not since litigation and lawyers took over afternoon TV. Somebody has to be to blame. We don't have Road Traffic Accidents any more – we have Road Traffic *Incidents*. Strokes are no longer cerebrovascular accidents, they are cerebrovascular *incidents*.

I should therefore call this section Incidental Death.

Today you might spill your coffee by incident or incidentally put too much cold water in the bath. Just think of all the incidental pregnancies happening across the world at the moment. I suppose in the sense of a coming together, that might not be so inaccurate.

It annoys me (this saddle is beginning to chafe). People have accidents. Things just happen. Really. It is not the tree's fault that it fell on someone or the wind's or God's and certainly not the fault of the dead guy who just happened to be passing at the time. Maybe he should have been prepared for this eventuality and left home later or earlier or…

I could concede to unlikely coincidence. I could call this section Unlikely Coincidence Death. Is the banana skin, recklessly discarded, the person who threw it down or the idiot who didn't watch where he was walking, the blameable quantity? Or did those three events happen to coincide to cause a greater, more unfortunate event? Then again, I'm not a solicitor clawing money from the fractured spine and twisted ankle of every victim of circumstance. Circumstance. Circumstantial Death? How does that sound? That too makes sense. After all, most 'accidents' are merely a series of circumstances. Circumstantial Death it is then, except…

This could go on for ever. It is just semantics. Semantics for lawyers and overpaid union leaders and HSE executives. Whether something happens as a result of a series of other events, due to circumstances beyond anyone's control or by coincidence, *it happens.*

The fact is though, that these deaths happen and, without wanting to apportion blame in any way, many of them are preventable.

A quick search for accidental deaths (yes, I'm sticking to my guns), gave me a list of 15 possibilities[141]. I'm going to use that list as a jumping off point (sorry).

---

[141] https://likes.com/weird/the-15-most-common-accidental-deaths

1. Falling Objects

Among these were death by falling sacrificial goat (it fell from a church roof), crushed to death by a huge Taco Bell sign, killed by an air conditioning unit that fell from the 7th floor, killed by a flying lawnmower, died after a coconut fell on his head (about 150 people die a year from coconut madness) and a soccer fan who was killed by a flying toilet. There's about a 1 in twenty million chance of being hit by a falling tree, according to the Health and Safety Executive.

2. Drowning

About 372,000 people a year die by drowning[142] and twice as many males as females are likely to drown. It is suggested that it is more common among men because a) they are more likely to venture into water and b) are more likely to take a watery risk by combining the event with alcohol or drugs or swimming alone and getting into trouble. I would suggest as c), men are clearly more stupid.

75% of the drownings in the world are due to floods. The tsunami on Boxing Day of 2004 in south-east Asia managed to add the names of about 300,000 to the list. That is a phenomenal amount. 275,000 people were killed in fourteen countries across two continents, with the last two fatalities being swept out to sea in South Africa, more than twelve hours after the earthquake[143].

In Britain in 2013, there were 381 drownings (or water-related deaths)[144]. 163 died off the coast of Britain in 2015 according to the RNLI.

Interestingly, your socioeconomic status also increases (at the lower end) your chances of drowning, as does being a member

---

142 www.who.int

143 http://www.thebcom.org/ourwork/reliefwork/96-the-boxing-day-tsunami-facts-and-figures.html?showall=1

144 www.rospa.com

of an ethnic minority - possibly due to differences in opportunities to learn to swim and/or a lack of higher education - in Bangladesh, children whose mothers have only primary education are at significantly greater risk of drowning compared with children whose mothers have secondary or higher education:

> 'Duration of education is often used as a proxy for social background in many studies which assess the socioeconomic differentials in injury risk. Although the existing data is occasionally controversial, there is evidence to suggest that higher parental education leads to higher levels of awareness of the existing environmental risks for their children and to the development of appropriate compensatory mechanisms. Therefore, where drowning and near drowning incidents are concerned, one should always have in mind the role that higher education, better occupation, and higher income can have on drowning risk.[145]'

And finally, there are those from the Boondocks – presumably because all they see is a sea of corn and don't know how to handle water over the height of their toes.

According to WHO though, much of it is down to common sense (my words, not theirs) when the risk factors are examined:

> 'infants left unsupervised or alone with another child around water;
> alcohol use, near or in the water;
> medical conditions, such as epilepsy;
> tourists unfamiliar with local water risks and features.'

---

[145] *Risk Factors for Drowning and Near Drowning Injuries* Eleni Petridou, MD, MPH

The last of the drowning stack of stats is of those who travel on water. There is a reason why we don't have gills and why we tend to keep something buoyant between us and the depths. We shouldn't be there. It's the same with aeroplanes. How more unnatural can one act be than to voluntarily climb into a metal can, which is surrounded by 50,000 gallons of flammable liquids, then allow it to propel you into the air and then trust it to keep you in the air until, wait for it, you come down for a *controlled* landing? Never mind the mountains, the storms, the engine-bound birds, the other planes, the UFOs and God knows what else that can bring you down between A and B.

As for boaty stuff - The QE2 held 4,800 gallons of fuel. That was a coracle compared to today's ships. Freedom of the Seas holds 1,097,021 gallons[146]. Why would you want to be in the middle of the Atlantic smoking a fag on top of what is essentially a highly flammable nuclear weapon? If you don't burn to death, you will drown – in bits.

I digress. I don't like risks. If I think about these things too much, I get scared.

That wasn't the last of the drowning facts. I'm just going to send you cheerfully onto Alcohol Poisoning with a little pot-pourri of facts from www.ilsf.org:

- One quarter of drowning victims were swimmers. This is discomforting. I thought I was safe.
- Young children 2 to 4 years of age have a higher risk of drowning than any other age group. Most of these children are alone and playing near water when they fell in and drowned. The backyard swimming pool is the riskiest site for these youngsters.
- Four out of every ten drownings happen within two meters of shore or the pool

[146] www.cruiselinefans.com. Bless 'em.

side. And one-quarter happen in shallow water one meter deep or less. They say the same about shark attacks. It just gets better.

- One-third of water-related deaths occur after dark, including fatal boating collisions.
- In Finland e.g. yearly over 50 accidents occur while driving over ice that is too thin to support the car. Why would anybody drive over ice in the first place? Go *around* the ice, nutters!
- 20% of all drownings occur at private homes.
- Few victims in boating deaths were wearing a life jacket.

Comforting, eh?

3. Alcohol poisoning

We tend not to take this too seriously. Drinking excessively is a part of growing up which then becomes a part of life. Sometimes it becomes a habit. Sometimes a need.

Although this section is about the sudden imbibing of massive amounts of alcohol to the point where even the pink elephants end up in A&E, I think it's fun and informative to release into the wild a few facts, just to see if anyone in their Audi Titty can drunkenly mow them down and take them home for an inebriate-appropriate snack between two large pieces of stale bread while urinating unawares upon the sofa.

Alcohol stops you sleeping – this is a surprise. Every time I've had too much to drink, I have slept like a top. I haven't woken up so easily, but isn't that the point of days off? Recovery?

Why does it stop you sleeping? Because you slip too quickly into what is known as deep sleep. Normally we go through stages to get to deep sleep, but a good half dozen pints gets you on the sleeping equivalent of Concorde.

The problem is, you then miss out on the pleasure of the tonic/clonic fall out of bed where you hope the wife didn't see you shudder like a sheep on a snowy hill and we miss, more importantly, REM sleep. Because you go so quickly into deep sleep, you also come out of it quickly, rise into REM sleep and leap beer belly first into wakefulness. The body hasn't had that period of time, in deep sleep, to recover from the effects of the previous day and the bevvies on top of that. This is why you wake up knackered (at the very least). We normally have about seven REM cycles a night[147] – with the alcohol this could be as little as two cycles, which shortens considerably the body's chance of a full recovery.

As I might have hinted at earlier, in my quiet way, I have recently lost weight – two stone in two months. How? I stopped drinking. I was drinking too much, much too much, and I felt sluggish, tired, out of breath. When I looked in the mirror, I would get into arguments with the fat bastard staring back at me.

Alcoholic drinks account for 10% of 29 to 64-year olds in the UK's daily intake of added sugar. That is a lot. A pint of cider has about five teaspoons of sugar in it. That's healthier than Coke, I hear you scream. Should I drink Coke? Is that what you want? No. You will bloat like a puffer fish if you drink those revolting fizzy soft drinks. Long term use of alcohol can also lead to diabetes because it messes with the body's head (if that makes sense). The body is like, 'Oh, wow, the guy's pumping sugar into me again' and it works like a Trojan to get rid of what it effectively sees as a poison, the excess sugar. This over time, can reduce the effectiveness of insulin, which leads to high blood sugar levels. If you do become diabetic and you still drink, you are then but a sip away from a hypoglycaemic coma. All those diabetic meds don't work properly anymore. Time to order, at the very least, a shiny new metal leg, at worst, a nice, burnished coffin.

Then of course, there's the old favourite – liver disease.

---

[147] www.drinkaware.co.uk

The liver carries out hundreds of functions, but about twelve main ones. Here is one of those lists that some people love and towards which others just wave a dismissive hand. I'll let you know, lists got me through my nursing exams.

1. Regulation of blood glucose/storage and release of glucose as needed – see above.
2. Blood clotting – the liver helps control the production of proteins, including one called Albumin (like the egg white) and, among many other things, Vitamin K, all of which stop you bleeding to death.
3. Temperature regulation – if you get too hot, you explode. Like a roadside badger.
4. It breaks down fats – this helps produce energy and ketones, which in turn are a source of energy for the muscles and tissues. If it doesn't do this, see 3.
5. It helps to synthesise vitamins – no vitamins, no life.
6. It breaks down drugs – Poisons in the body are metabolised by the liver, meaning that the liver chews them up and spits them out once they have done their job. Excess goes into the urine or faeces.
7. It helps make bile salts – these help the body absorb and digest the fats and break them down (emulsify). Enzymes in the pancreas can then help the fat do what the fat has to do – no bile salts, no life and very smelly poo before you croak.
8. It helps make blood cells in various ways. No blood, no…you know this one.
9. It helps to clean the blood.
10. Production of cholesterol to help carry fats through the body.
11. Conversion of harmful ammonia to urea – proteins are broken down into urea and peed out.
12. The liver processes all the blood that goes through the stomach and intestines.

When your liver goes, so do you. That's just after your

eyes and skin turn glow-in-the-dark-yellow due to all the crap it can no longer remove from your blood stream. This is called jaundice.

Now we are venturing into the dark underbelly of the Anatomy Continent, the places where things go wrong, where not even Stanley or Livingstone, nay, even Indiana Jones, would dare to tread.

Eventually, and for various reasons that are not always alcohol related, but alcohol is the major culprit, the liver might become cirrhotic; it will develop cirrhosis. This means that, due to scarring, that beautiful, moist, pink, three pounds of fleshy factory becomes a wizened piece of sun-baked roadkill. The liver dies and, usually, so does the owner.

This is long-term alcohol poisoning. This is what happens if you survive the binges. But what happens to cause that cataclysmic, once in a lifetime trip to the Darklands that results in your demise?

Many of the things we put into our body are poisons. It is the amount that goes in, over what period of time and the body's ability to cope with it that counts. The body can only cope with about a unit of alcohol an hour. Any more than this and the levels become toxic.

The symptoms of acute alcoholic poisoning are:

- Confusion – because it has a toxic effect on the central nervous system. If you look at alcoholics, you can see that sometimes they have tremors, an unsteady gait or slurred speech. Ozzy Osbourne is a good example.
- Loss of coordination
- Vomiting – causing the affected to choke on their vomit or leading to fatal lung damage or pneumonia
- Seizures – due to lowered blood sugar levels.

- Irregular or slow breathing (less than eight breaths a minute) - stop breathing, causing the affected to die.
- Blue-tinged or pale skin – a sure sign of impending doom.
- Low body temperature (hypothermia)
- Stupor - when someone's conscious but unresponsive
- Unconsciousness - passing out[148]

If you look at each of these symptoms logically, then it doesn't take much to see where the old Jimi Hendrix situation comes to pass.

On top of all this, there's still room to become severely dehydrated, which can lead to permanent brain damage. That's if you survive the heart attack, of course.

So, there you go. That's alcoholic poisoning for you.

Did I mention that I have recently lost weight?

4. Roller coasters

I have no idea why this is on the list. There are apparently about four deaths a year from the use of rollercoasters. My inclination is that it is old farts like me who should have known better, having a heart attack on the three-hundred-foot vertical drop while trying to catch my glasses which seem to have developed a new gravity field of their own.

However, this is on the list for a reason and so is probably worth investigating.

Overall, there were 52 deaths related to amusement park rides in general between 1990 and 2004 based upon a 2005 report by the Consumer Product Safety Commission[149]

Unlike the alcohol poisoning section, I'm not going to go

---

[148] www.drinkaware.co.uk

[149] www.inquisitr.com July 9, 2014

into too much detail. There's only so much you can add to 'People Die Due To Human Error'. However, like boats and planes and even trains, I think we should be aware of the risks we are taking. I did go on a rollercoaster at Disneyland Paris – the Aerosmith one. I screamed like a girl. Like…a…girl.

Here are some examples from an article that the Daily Mirror did in 2015. If that doesn't cease stupidity, nothing will.

> 'Space Journey. Eco-Adventure Valley Space Journey at Overseas Chinese Town East, Shenzhen, China.
>
> This rocket simulator ride had a multi-car centrifuge that whirled inside a domed screen showing movies about space. On June 29, 2010 one of the cars came loose and lost all power. The dome plunged 35 feet to the ground, on fire, with more than 44 people inside. Six were killed and ten more taken to hospital unconscious. Reports say the tragedy was caused by a faulty, loose screw.
>
> Battersea Big-Dipper
>
> The worst roller coaster tragedy in history happened in May 1972, on the old wooden roller coaster. The rope that hauls the cars to the top of the launch hill snapped and then, the anti-rollback mechanism failed. It caused a chain of cars to hurtle back into the boarding area, smashing into a wall and killing five children and injuring 13 more.
>
> Mindbender rollercoaster at Galaxyland, West Edmonton Mall.
>
> Mindbender - the world's largest indoor triple loop roller coaster – was dubbed one of the "safest rides in the world" when it was built. But in June 1986

missing bolts caused the four-car-train to fly from the track. The last car hit support structures before crashing into a concrete pillar in front of a packed audience watching a concert. Three people fell to their deaths and the fourth suffered serious neck and chest injuries.

Fujin Raijin II

The thriving amusement park opened in 1970 with more than 40 rides and attractions. But on May 5, 2007 the Fujin Raijin II roller coaster was derailed, sending cars flying off the tracks.
A 19-year-old university student was killed, and 19 other people were injured. An investigation showed a broken axle caused the derailment, and none of the ride's axles had been replaced for 15 years.'[150]

Is it just me? Let's look at these again: 'a faulty, loose screw', 'a rope snapped and the anti-roll-back mechanism failed', 'missing bolts' and 'showed a broken axle…none of the ride's axles had been replaced for fifteen years'. Maybe those people who say that there is no such thing as an accident have a point here. Someone was sure as hell to blame with these.

5. Diving

Again, not sure why this is on the list. It's aimed at those who dive, as in from a diving board. However, I want to look at scuba diving too. It's all diving, after all.

'9 fatal incidents occurred in the UK during the 2015 incident year. This is the lowest number for over 20 years (in 1983 there were 8 recorded

---

[150] www.mirror.co.uk/news/uk-news/alton-towers-rollercoaster-crash-worlds-5822862 4 Jun 2015

fatalities). The nine fatalities in 2015 involved divers aged between 44 and 59 with an average age of 52.

- Five of these cases involved the casualty falling unconscious under the water. In all these incidents, where a casualty falls unconscious underwater, the rescue becomes much more problematic.
- Three confirmed cases involved divers who suffered a 'nondiving' related medical incident (for example a heart attack) whilst in the water. There are two additional cases where it seems very likely that the diver suffered a 'medical event' whilst underwater, although evidence to substantiate this assumption is not currently available.
- Five cases involved divers diving in a group of three. Diving in groups of three (or more) brings additional complexity to a dive and can generate problems that don't exist with pair diving. However, it is not clear whether trio diving directly contributed to these fatalities. BSAC recognises that, at times, it is necessary to dive in a group of three.
- Two cases involved a rapid ascent whilst carrying out an alternate source ascent. These fast ascents may not have directly contributed to the fatality.
- Four cases involved a separation of some kind and three of these separations happened in cases where more than two divers were diving together.
- There was one case of a solo diver, diving

> on a rebreather, where insufficient information is available to understand the cause of the incident.'[151]

In a report (June 2007), the Divers' Alert Network stated that

> 'the initial triggering event (which began the sequence of events that ultimately led to death) of scuba diving fatalities for the year were as follows:
>
> •Insufficient Gas: 14%
> •Rough Seas/Strong Current: 10%
> •Natural Disease: 9%
> •Entrapment: 9%
> •Equipment Problems: 8%
> •Could Not be Determined: 20%
>
> Note that many of these events could/should have been avoided - such as running out of gas. This fact holds true for many of the deaths reported each year.'

The advice that you should 'avoid running out of gas' is perhaps a trifle obvious. You would think. Maybe they should have added 'didn't leave enough time between finishing his sandwiches and pop'. My mother always said that this was fatal.

If I was scuba diving, which is as likely as me bungee jumping without the elastic band, I would superglue my eyes to the gasometer or whatever it is. The fact that 20% of deaths, that's one in five, could not be determined is alarming. If one in five people in your street died overnight you would, quite rightly, want to know why.

As for the less deep and less appliance dependent diving-from-a-board type of diving, there are apparently 6,700 people a

---

[151] NDC Diving Incidents Report - 2015

year hospitalised due to diving accidents/incidents - 'some of them even lost their lives'.

Research-wise, this is a pretty weak area. People only seem to care about scuba incidents. I can't relate to that. I hit my head against the bottom of a pool when I was young. I felt my neck contort like a knotted swan. Now that I can relate to. I'd like to know more. Approximately 3,000 people world-wide become partially or completely paralysed each year as a result of breaking their necks. Most of these injuries occur while diving into shallow water[152]. I think that is relevant.

Should we have a quick peek at that other, ridiculous, nonsensical, life-mocking, unnatural, unfeathered, unwinged form of diving – skydiving?

Yeah. We have time.

And let's not beat about the bush here – this is called, in the real world, falling. I don't care what Buzz Lightyear says, falling with style is still falling.

> 'In 2015, USPA recorded 21 fatal skydiving accidents in the U.S. out of roughly 4.2 million jumps…In the 1970s, the sport averaged 42.5 skydiving fatalities per year. Since then, the average has dropped each decade. In the 1980s, the average was 34.1; in the 1990s, the average was 32.3, and in the first decade of the new millennium (2000-2009), the average dropped again to 25.8. Over the past six years, the annual average continues its decline to 22.3… 2015 had 0.56 fatalities per thousand USPA members. And estimating about 4.2 million jumps last year, that's one fatality per 200,532 skydives.[153]'

If I did it, guess who would be that 200,533rd? Yep, you

---

[152] www.ilsf.org

[153] http://www.uspa.org/facts-faqs/safety

guessed it.

This is what I understand by skydiving statistics:

Stuart Morris, from Skydive North West, gives a slightly different account of the figure:

> "The risk of fatality from skydiving is thought to be somewhere in the region of 1 in 100,000 skydives. That means, for every 100,000 skydives made – approximately 1 person dies engaged in the sport.'

That's less than half as much chance of surviving as the USPA gives you. In his piece, Stuart says that he has done 3,500 skydives. By his figures that leaves him 96,500 more attempts until…you know.

In an effort to blind fearful cynics like me, www.skydivemag.com came up with an article about skydiving involving the micromort – that is 'the risk out of a million of dying from a particular event'. Skydiving, it claims, is at 10 micromorts. It says that this is the same risk as having a general anaesthetic.

> 'In 2002, anesthesiologist Dr Robert S. Lagasse of the Albert Einstein College of Medicine in New York published a study in *Anesthesiology*, the specialty's leading journal, which challenged the Institute of Medicine report.
>
> Lagasse presented data on surgical mortality from two academic New York hospitals between the years 1992 and 1999. Deaths related to anesthesia errors were…only 14 deaths out of 184,472 surgeries–a rate of 1 death per 13,176 cases. However, Lagasse's anesthesia-related mortality rate of 1 per 13,176 surgeries was significantly different that the Institute of Medicine's rate of 1 death per 200,000-300,000 surgeries.'[154]

---

[154] https://theanesthesiaconsultant.com/anesthesia-facts-for-non-medical-people-how-safe-is-anesthesia-in-the-21st-century/

In 2011, Lou Davis, a Clinical Nurse Educator said on www.quora.com, 'Most recent studies put the mortality rate at 1: 100,000 anaesthetic administrations'.

Well, there you go, Stuart was, even on a bad day for anaesthesiologists, inaccurate, if you believe theanesthesiaconsultant.com and the Institute of Medicine's figures. On the other hand, 1 in 13,176 is fairly shitty odds and Stuart is rendered very much correctomundo. At 10 in a million chance, the Skydiving.com estimate, that's one in every one hundred thousand. It turns out that, all in all, I stand pretty much the same chance of dying from both.

I have no plans to do either at the moment. If I had to toss a coin? I'd rather go and stand in a field with a metal umbrella during a lightning storm. How's that for a segue?

6. Lightning

> 'Lightning is an electric charge that travels through the air about 270,000 miles per hour, or 75 miles a second, and typically lasts for one 10,000th of a second. During its brief existence it burns at 30,000°C, or six times hotter than the Sun.[155]'

24,000 people are killed by lightning worldwide each year.
Stay indoors.
Seriously.
This from the US National Weather Service:

Odds Of Becoming A Lightning Victim (based on averages for 2006-2015)	
Estimated US pop. as of 2016	318,000,000
Average number of deaths reported	31

[155] www.independent.co.uk 6 July 2015

Estimated number of injuries	279
Odds of being struck in a given year	1/1,042,000
Odds of being struck in life time (estimated at 80 years)	1/13,000
Odds you will be affected by someone struck (10 people for every one struck)	1/1,300

Although it's tempting to wonder if 'odds of being affected by someone struck' could be looked at as any sort of X-Men or superhero contact, I think it probably just means anyone you know who got hit by lightning.

I don't know anyone who has been struck by lightning. Maybe the odds in the UK are less because we are so tiny and surrounded by sea and pylons. The stats available suggest that you are more likely to get hit by lightning than win the lottery. Neither has happened to me.

I'll let you know.

7. Machinery

This is where we come back to reality and enter the realms of well-kept statistics. That is a relief. Although this category is one of machinery, I'm twisting it slightly by including accidents and deaths at work as well.

Britain is an amazingly safe place to be. We have come from being 19th century child-killers-for-profit to one of the most safety-conscious countries in the world, especially when it comes to the workplace, governed primarily by the Health and Safety at Work Act 1974 and its subsequent amendments.

As I live in South Yorkshire, I'm going to use as an example the coalmines[156].

---

[156] Image: www.theguardian.com/politics/reality-check-with-polly-curtis/2011/sep/28/reality-check-how-dangerous-is-mining

The mines have been a part of the social landscape up here since the mid-1300s when Sir John Fitzwilliam gave permission in 1367 for mining to take place on his estate near Elsecar, south of Barnsley. It lasted, as anyone up here will tell you, until it was euthanised in the mid-eighties and then slowly dribbled out of existence thanks to cheap imports, a change in the use of fossil fuels and a desire by successive governments (of all colours) to get rid of the expense and dissolve the unions affiliated to the mines and/or any representative body of worker's rights.

[157]**Deaths of coalminers since 1700**	
1700-1750	93
1750-1800	267
1800-1850	3486
1850-1900	59580
1900-1950	84331
1950-2000	16599
Total	164356

It is difficult to get an umbrella figure for all deaths in all industries prior to the late 20th/early 21st centuries. The HSE, as we shall see, keeps excellent figures for modern industry, but there was sparse information available prior to this without breaking the industries down.

For the mining information, I have used information gathered mostly from one source, www.healeyhero.co.uk., run by Alan Beales, who worked at Gedling Colliery for thirty-five years and has spent a considerable amount of time researching fatalities in the coalfields.

The names and mode of death have been meticulously noted by Mr Beales and it is from that that I will give you some of the awful ways in which these people perished:

1) Knocked off a platform, drowned in the sump

[157] The Coalmining History Resource Centre

The sump of a mine was the lowest part into which water could drain. 'The bottom of a shaft, or any other place in a mine, that is used as a collecting point for drainage water.158' So essentially, this guy was knocked from a platform (no safety rails, this was 1865) and then fell into a drainage area and drowned. Whether he drowned because he couldn't swim or because he had sustained some other injury which rendered him unconscious or unable to swim is unknown. Either way, in the cold and the darkness of 1865, it must have been an appalling way to go.

2) Crushed by wagons, on the surface.

Here is a description of a death from 1924 in Conisbrough's Cadeby Colliery[159]:

> 'The wagon being filled was in the "hopper" (a container in which coal is stored in readiness for dispensation). Two wagons were standing lower down a few yards away. It was quite daylight. When he had filled the wagon he took the "scotch" (the "scotch" was a block of wood across the rails) off to lower the wagon. He looked on both sides before doing so. When he lifted the brake off, the wagon ran down to the two wagons. He heard a shout. "Oh!" and that was all. He looked underneath the wagon and saw Hornsby lying in the centre of the rails. Opposite was a cabin used by platelayers. He did not see Hornsby when he lowered the wagon...The Coroner said it was evident the wagon had run away from them. Had the brakes been in working order probably the

---

[158] www.caseyresearch.com/resource-dictionary/definition/sump

[159] http://conisbroughanddenabyhistory.org.uk/article/fatal-accident-at-cadeby-colliery-elderly-labourer-crushed/

> accident would not have occurred...A verdict of "Accidental death" was returned.'

How can someone get crushed by two wagons, in daylight? I have seen bus drivers nearly turned to jam between two buses at the depot. We all blink.

3) Fell down the shaft.

> 'Falkirk Herald 12 January 1871
>
> Fearful Fall - On Friday last a pit-head man named Alexander Gray, in the employment of the Redding Colliery Coy., near Falkirk, fell down one of the shafts of No. 15 Pit, on Green Wells Farm, and was instantaneously killed. The deceased at the time of the accident was engaged pushing an empty waggon towards the pit-mouth to place it on the cage which he seemingly expected to be there, but without, it is supposed, looking to discover this, he ran the hutch right forward, and the cage not being at the mouth of the pit at all, he and the hutch disappeared in the shaft. Two men at once descended and brought the lifeless body of the unfortunate man to the pithead, whence it was conveyed home. The deceased was thirty-five years of age.'
>
> 'Scotsman 19 January 1871
>
> On Tuesday morning, the body of a man, dreadfully mangled, was found in one of Messrs Wallace's pits at Neilston Kilsyth. The unfortunate man was David Tait, a nailer, and had been in the town the previous day seeking for work. It is supposed that he went to the pit in search of a

night's shelter and accidentally fell down.'[160]'

I think in modern times we think that anyone falling down a pit shaft must, frankly, be a farthing short of a penny. The examples above suggest this not to be the case. We are, thankfully, mollycoddled these days by warning signs and railings and Health and Safety Regulations and a thing called the electric light. They had none of these.

4) Suffocated by afterdamp, following an underground fire.

> 'Afterdamp is the toxic mixture of gases left in a mine following an explosion caused by firedamp, flammable gas found in coal mines. Firedamp is the name given to a number of flammable gases, especially methane. The gas accumulates in pockets in the coal and adjacent strata, and when they are penetrated, the release can trigger explosions. Historically, if such a pocket was highly pressurized, it was termed a "bag of foulness"[161]. Firedamp itself can initiate a much larger explosion of coal dust[162]. It consists of carbon dioxide, carbon monoxide and nitrogen. Hydrogen sulfide, another highly toxic gas, may also be present. However, it is the high content of carbon monoxide which kills by depriving victims of oxygen by combining preferentially with haemoglobin in the blood…Afterdamp was the deadly gas which caused the majority of casualties in the many pit disasters of the British coalfields.'

---

[160] www.scottishmining.co.uk/350.html

[161] *Bag of foulness.* A Glossary of Terms Used in Coal Mining. William Stukeley Gresly (1882). London: E. & F.N. Spon.

[162] *A Glossary of Mining and Metallurgical Terms.* Easton, Pennsylvania: American Institute of Mining Engineers. 1881.

> '8 men were killed by afterdamp and 66 escaped, but 1 of these died later from Injuries sustained while getting on the cage. The explosion originated in No. 5 mine, and there the heat and violence were so great that few of the 172 men in those workings could have lived any considerable time after the explosion. About 10 minutes after the first explosion, a second but less violent explosion occurred, carrying debris out of No. 6 shaft. The first and more violent explosion, accompanied by flame, carried timbers and quantities of mud up both shafts and blew off the explosion doors of the fanhouse at No. 5 shaft...The explosion wave in No. 5 mine traveling toward No. 6 shaft blew a large quantity of water from a depression near the shaft up the shaft. This quenched the flame and prevented it from entering No. 6 mine.[163]'

5) Fell out of the cage.

The cages in which miners ascended from or descended to the pit were open. In one instance a miner was hit by debris and knocked from the cage, to fall to his death.

6) Run over by a locomotive, on the surface. Leg amputated. Toxaemia
7) Injured thumb, death from Septicaemia.

I'm putting these two on a par because they both ended with the same result from a similar cause. We tend to take antibiotics and medical care for granted nowadays, but, as was seen, a simple injury to the thumb, which today would be treated by a quick visit to A&E and some tablets, then proved to be fatal when the wound became infected and the affected man died from septicaemia, a generalised infection that leads to the major organs shutting down.

---

[163] http://digital.library.unt.edu/ark:/67531/metadc12740/m1/79/

Toxaemia is slightly different in that, as this is connected to an amputation, I would imagine that it is connected to a crush injury. These systemic effects are caused by a traumatic break down in the skeletal muscles, called rhabdomyolysis (also known as 'crush syndrome'). As muscle cells die, they absorb sodium, water and calcium; the rhabdomyolysis releases potassium, myoglobin, phosphate, thromboplastin, creatine and creatine kinase – which basically leads to a massive chemical imbalance. We have learned that there is a strict compliance to the control of bleeding in crush injuries which, if not followed, can lead to the poisons being released into the body and the victim succumbing to heart attack or, later, renal failure. I once attended a lecture by a fireman about this. He was most persuasive in his terminology.

8) Hit by a prop, death from cancer due to injury.

This is an interesting one. My mother swore that you could die from cancer after an injury or that surgery on cancer, far from healing it, encouraged it to run rampant through the body. 'People often think that a knock or injury to a bone can cause a cancer. But research studies do not support this. It is more likely that an injury causes swelling, which shows up a cancer that is already there.[164]' The gentleman concerned, working in the environment that he did, leading the lifestyle that he did, might already have had the disease.

9) Kicked by a pony, death from peritonitis.

Once again, the peritonitis wouldn't have had the chance to set in these days and if it had, then antibiotics would have taken care of it.

10) Fall of roof

This isn't a typo. It's when the roof falls in, not 'fall off roof', as I

---

164 www.cancerresearchuk.org

first read it in bemused fashion. The Mine Safety and Health Administration in the USA stated that between 1976 and 1985, there were 1100 workers killed in mining accidents, 374 of which were blamed on roof collapses. In the Crandall Canyon coalmine collapse of 2007, 6 miners and 3 of their rescuers were killed. As a seismic event, the event had a magnitude of 3.9.

> 'The scientists reported that of the 17,000 or so earthquakes detected in the Wasatch Plateau region of Utah from 1978 to 2007, 98% of them were attributable to mining activities, not natural seismic activity. So the odds were very clearly against the mine collapse resulting from an earthquake. Even without context, this is an amazing statistic: in this coal region, the seismic signal from human activity completely drowns out the earth's own perturbations…Miners were primarily extracting coal from Crandall Canyon by the room-and-pillar method, in which continuous miners [the machines used to cut the coal from the face] cut large swaths of coal. The problem, for mining companies, is that huge pillars of coal have to be left intact to support the 1,500 feet of rock soil sitting on the roof of the mine. That leads some mine owners to engage in so-called retreat mining, in which the pillars holding up the roof are partially removed. The process accounts for 10% of underground coal production, but it's easy to understand how it is the most hazardous coal mining activity. And that's exactly what the Crandall Canyon workers were engaged in when the mine collapsed. After the collapse began, seismic waves traveled [sic] east across the mine and the roof began to collapse across up to 50 acres of mine. [165]'

---

165 *The Autopsy of a Coal Mine Collapse.* www.wired.com/2008/06/the-autopsy-of/

11) Explosion of methane.

Dotted around Barnsley and other old mining areas are little chimneys that stick out of the ground. These are to let the methane gas escape from below ground. If they weren't there, there would be explosions of mind-boggling size.
This is from the CDC website:

> 'Methane explosions occur when a build-up of methane gas contacts a heat source and there is not enough air to dilute the gas level below its explosion point. Likewise, fine particles of coal dust in the right concentration that contact a source of heat can also be explosive. Hybrid explosions consisting of a combination of methane and coal dust can also occur. Methane is formed as a by-product of the coal formation. The methane that is absorbed in the coal is released as the coal is mined or it migrates from surrounding sources above or below the coal seam through fractures created by the coal extraction process.'

12) Drawn into the Whim Gin wheel.

A whim, also called a whim gin or a horse capstan, is a device used in mining for hauling materials to the surface. The example given was from 1838, at Awsworth Colliery.

The idea that you could somehow be drawn/dragged into machinery is horrendous. You would have felt every second of your death.

13) Pneumoconiosis

> 'The term "pneumoconiosis" refers to a group of lung diseases caused by the inhalation, and

retention in the lungs, of dusts. The disease is characterised by scarring and inflammation of the lung tissue. It is an irreversible condition with no cure. Symptoms include, shortness of breath, persistent cough, fatigue, laboured and rapid breathing, chest pain. These can seriously affect normal daily activity and lead to various complications which can be fatal.[166]'

Coal workers' (or 'worker's, for the pedants out there) pneumoconiosis – also known as black lung - is a lung disease caused by deposits of coal dust in the lungs. Eventually, the miners get so much coal dust in their lungs that the body is no longer able to expel it in the mucus or through the lymphatic system. This over time will fibrose (thicken the tissues by scarring and take away the elasticity) the lungs and impede the ability to clear the chest and to breathe. This leads to emphysema or Chronic Obstructive Pulmonary Disease. It is irreversible and can only be treated symptomatically.

It is a debilitating disease that affects every aspect of the individual's life – health, income, relationships, quality of life, sleep, eating, walking – there isn't an aspect of life that such a disease does not attack.

It's not just inhaled coal dust that contributes to lung disease. I nursed a patient who was a gardener for the council and had contracted COPD from the chemicals he had used in the course of his work. There is also mesothelioma, a rapidly spreading cancer caused by the inhalation of asbestos. The annual number of mesothelioma deaths has increased from 153 in 1968 to 2,291 in 2011[167] and there were 2,567 in 2014[168]. They are expected to peak around 2020.

Other industrial lung diseases include, silicosis,

---

166 *Pneumoconiosis (excluding asbestosis) in Great Britain 2014* – Health and Safety Executive.

167 Asbestos related disease statistics; Health and Safety Executive (HSE)

168 www.cancerresearchuk.org

pulmonary siderosis and berylliosis. With more attention paid to working practices, reinforced my legislation and monitoring by unbiased agencies, it is anticipated that numbers will go down. On the other hand, who knows what we will invent tomorrow in order to do ourselves in?

In modern times, of course, we have the Health and Safety executive keeping an eye on the workplace and we are inundated with laws laid down to protect us in the workplace, both physically and mentally.

Another important factor in our protection, and I hate that I am saying this, is litigation. Those afternoon adverts on the cheap and nasty channels are no more than ambulance-chasing, mostly done by bad actors who failed the audition for the porn movie that morning. But, if employers are scared that they might get sued, then any protection that the worker gains has to be a good thing. Some employers remain shockingly blasé about their workers' psychological and physical safety.

Industry	**Employee**	**Self-Employed**	**Workers**	**Members Of The Public**	**Total Fatal Injuries**
Agriculture	10	17	27	2	29
Mining/Quarrying	2	-	2	-	2
Manufacturing	27	2	27	-	27
[169]**Utilities**	8	-	8	-	8
Construction	27	16	43	2	45
Services	33	4	37	99	136
All Industries[170]	105	39	144	103	247

---

[169] Gas, Electricity, Water, Sewerage, Waste, Recycling

[170] Number of fatal injuries by main industry 2015/16 - from Statistics on Fatal Injuries in the Workplace in Great Britain 2016. HSE.

If we look at the chart above, from *Statistics on Fatal Injuries in the Workplace in Great Britain 2016* by the HSE, what is startling is how *few* accidents we are having nowadays. And we fare quite well next to our continental cousins.

In the USA in 1900 there were 1,489 (recorded) deaths among miners out of 448,551 employees. In 2015 there were 11 out of 102,804. That's 0.33% and 0.01% respectively. That's a reduction by 33 times.

Incidentally, in England, The Oaks mine explosion remains the worst mining accident in England, claiming 388 lives on 12 December 1866, near Barnsley in South Yorkshire.

Here are a few of the laws made for our protection in the workplace.

1) Management of Health and Safety at Work Regulations 1999
2) Workplace (Health, Safety and Welfare) Regulations 1992
3) Health and Safety (Display Screen Equipment) Regulations 1992
4) Personal Protective Equipment at Work Regulations 1992
5) Manual Handling Operations Regulations 1992
6) Health and Safety (First Aid) Regulations 1981
7) The Health and Safety Information for Employees Regulations 1989
8) Employers' Liability (Compulsory Insurance) Act 1969
9) Reporting of Injuries, Diseases and Dangerous Occurrences Regulations 1995 (RIDDOR)
10) Noise at Work Regulations 1989
11) Electricity at Work Regulations 1989
12) Control of Substances Hazardous to Health Regulations 2002 (COSHH)

Next time you roll your eyes (as my wife did this morning

when she made her way to Sheffield for compulsory Moving and Handling training), just think about how bad it used to be.

And this is the tip of a mighty, non-harmful iceberg. Thank God for it, I say.

INTERMISSION

Time for an accidental break.

This has got a bit heavy, wouldn't you say? I think we should have fun, so I've looked up a few of those classic, mad, insurance claims for you to read:

- 'I was driving along the motorway when the police pulled me over onto the hard shoulder. Unfortunately, I was in the middle lane and there was another car in the way.'
- 'Going to work at 7am this morning I drove out of my drive straight into a bus. The bus was 5 minutes early.'
- 'I was driving along when I saw two kangaroos copulating in the middle of the road causing me to ejaculate through the sun roof.
- 'The accident happened because I had one eye on the lorry in front, one eye on the pedestrian and the other on the car behind.'
- 'I started to slow down but the traffic was more stationary than I thought.'
- 'I pulled into a lay-by with smoke coming from under the hood. I realised the car was on fire so took my dog and smothered it with a blanket.'
- Q: Could either driver have done anything to avoid the accident? A: Travelled by bus?
- The claimant had collided with a cow. The questions and answers on the claim form were - Q: What warning was given by you? A: Horn. Q: What warning was given by the other party? A: Moo.
- 'I started to turn and it was at this point I noticed a camel and an elephant tethered at the verge. This distraction caused me to lose concentration and hit a bollard.'

- 'On approach to the traffic lights the car in front suddenly broke.'
- 'I was going at about 70 or 80 mph when my girlfriend on the pillion reached over and grabbed my testicles so I lost control.'
- 'I didn't think the speed limit applied after midnight'
- 'I knew the dog was possessive about the car but I would not have asked her to drive it if I had thought there was any risk.'
- Q: Do you engage in motorcycling, hunting or any other pastimes of a hazardous nature? A: 'I Watch the Lottery Show and listen to Terry Wogan.'
- 'First car stopped suddenly, second car hit first car and a haggis ran into the rear of second car.'
- 'Windscreen broken. Cause unknown. Probably Voodoo.'
- 'The car in front hit the pedestrian but he got up so I hit him again'
- 'I pulled away from the side of the road, glanced at my mother-in-law and headed over the embankment.'
- 'The other car collided with mine without giving warning of its intention.'
- 'I collided with a stationary truck coming the other way'
- 'A truck backed through my windshield into my wife's face'
- 'A pedestrian hit me and went under my car'
- 'In an attempt to kill a fly, I drove into a telephone pole.'
- 'I had been shopping for plants all day and was on my way home. As I reached an intersection a hedge

sprang up obscuring my vision and I did not see the other car.'

- 'I was on my way to the doctor with rear end trouble when my universal joint gave way causing me to have an accident.'
- 'An invisible car came out of nowhere, struck my car and vanished.'
- 'I was thrown from the car as it left the road. I was later found in a ditch by some stray cows.'
- 'Coming home I drove into the wrong house and collided with a tree I don't have.'
- 'I thought my window was down, but I found it was up when I put my head through it.'
- 'The guy was all over the road. I had to swerve a number of times before I hit him.'
- 'I had been driving for forty years when I fell asleep at the wheel and had an accident.'
- 'As I approached an intersection a sign suddenly appeared in a place where no stop sign had ever appeared before.'
- 'To avoid hitting the bumper of the car in front I struck a pedestrian.'
- 'My car was legally parked as it backed into another vehicle.'
- 'I told the police that I was not injured, but on removing my hat found that I had a fractured skull.'
- 'I was sure the old fellow would never make it to the other side of the road when I struck him.'
- 'The pedestrian had no idea which way to run as I ran over him.'
- 'I saw a slow moving, sad faced old gentleman as he bounced off the roof of my car.'

- ‘The indirect cause of the accident was a little guy in a small car with a big mouth.’
- ‘The telephone pole was approaching. I was attempting to swerve out of the way when I struck the front end.’
- ‘The gentleman behind me struck me on the backside. He then went to rest in a bush with just his rear end showing. ‘
- ‘I had been learning to drive with power steering. I turned the wheel to what I thought was enough and found myself in a different direction going the opposite way.’
- ‘I was backing my car out of the driveway in the usual manner, when it was struck by the other car in the same place it had been struck several times before.’
- ‘When I saw I could not avoid a collision I stepped on the gas and crashed into the other car.’
- ‘The accident happened when the right front door of a car came round the corner without giving a signal.’
- ‘No one was to blame for the accident but it would never have happened if the other driver had been alert.’
- ‘I was unable to stop in time and my car crashed into the other vehicle. The driver and passengers then left immediately for a vacation with injuries.’
- ‘The pedestrian ran for the pavement, but I got him.’
- ‘I saw her look at me twice. She appeared to be making slow progress when we met on impact.’
- ‘The accident occurred when I was attempting to bring my car out of a skid by steering it into the other vehicle.’
- ‘I bumped into a lamp-post which was obscured by human beings.’

- 'The accident was caused by me waving to the man I hit last week.'
- 'I knocked over a man; he admitted it was his fault for he had been knocked down before.'

END
OF
INTERMISSION

8. Medical malpractice

> 'In 1999, Americans learned that 98,000 people were dying every year from preventable errors in hospitals. That came from a widely touted analysis by the Institute of Medicine (IOM) called To Err Is Human. This was the "Silent Spring" of the health care world, grabbing headlines for revealing a serious and deadly problem that required policy and action.
>
> As it turns out, those were the good old days.'
>
> *Stunning News On Preventable Deaths In Hospitals* - Leah Binder. www.forbes.com. Sep 23, 2013

There is an assumption, reasonably so, that you will be safe in hospital or in the hands of your GP or any health specialist come to that. There is an unspoken, implicit bond of trust. Let us not forget though, that Harold Shipman accounted for 16.5% of all murders in England and Wales in 2002, a 14.8% increase on the previous year.

The same article goes on to say:

> 'According to a new study just out from the prestigious Journal of Patient Safety, four times as many people die from preventable medical errors than we thought, as many as 440,000 a year.'

Why? Is it because we are keeping better records? Is it because we have these hideous pasty-faced, grey-hearted lawyers ambulance-chasing 24-hours a day? Do we now know more because of the internet? Is it because we have become more accountable through law? Are we better protected by the law? Is the practice of whistle-blowing now more acceptable than in, say, 1999?

The article[171] upon which forbes.com bases its stats says that there are five areas for potential error

1. Errors of commission - This occurs when a mistaken action harms a patient either because it was the wrong action or it was the right action but performed improperly. When I was a nurse I did actually see the wrong arrow drawn in indelible ink on the wrong leg. Yes, it was for amputation.

2. Errors of omission - Errors of omission occur because of a failure to follow evidence-based guidelines. A simple example is that a patient on tablets such as Digoxin or Warfarin need regular blood tests to see that they are not overdosing. If you are on Warfarin and you find yourself exsanguinating on the High Street, you might well be overdue a blood test.

3. Errors of communication - Errors of communication can occur between 2 or more providers or between providers and patient. It is easily done in a moment of lost focus. Unfortunately, the error can sometimes be life-changing or fatal (which in itself is pretty life-changing).

4. Errors of context - Contextual errors occur when a physician fails to take into account unique constraints in a patient's life that could bear on successful, post-discharge treatment. For

---

[171] Journal of Patient Safety: September 2013 - Volume 9 – Issue 3. *A New, Evidence-based Estimate of Patient Harms Associated with Hospital Care*. James, John T. PhD

example, is the patient able to follow simple instructions about their medication or their exercise? There is no point telling someone with dementia to take their tablets three times a day.

5. Diagnostic errors - Diagnostic errors result in delayed treatment, the wrong treatment, or no effective treatment. A small subset of these might be included as errors of commission or omission. I suppose this is the most obvious one. It manages every once in a while to find headline space on a red top.

The Observer in September 2009 printed an article which stated that as many as one in six diagnoses were made in error.

> 'Between April 2008 and March 2009 there were 39,500 reports of incidents involving clinical assessment. Those included missed or wrong diagnosis but also related to scans that could have been misinterpreted or where the wrong body part was scanned or tests where patients' samples could have been mixed up.'
>
> Oops.

9. Poison gas

This is another area where we tend to be a little…casual. We take it for granted that the Corgi gas fitter was not in fact a dog and had done the job properly or that the pipes connecting our cooker to the mains will, despite the evidence, last forever. We all know that, once installed, household objects last forever.

A quick look at the stats from *Co-Gas Safety's Statistics on*

Deaths and Injuries[172] demonstrates that we are not as safe we assume ourselves to be. The good thing is that the numbers have decreased, thanks in part to tireless campaigning by people such as The Carbon Monoxide and Gas Safety Society and headlines such as those where a holiday company was found culpable in the deaths of children in a family holidaying abroad.

> 'It is well-known that carbon monoxide is colourless and odourless. It works by binding to red blood cells, which displaces oxygen and suffocates the body. Blood becomes 'sticky' and easily clots, which in some cases leads to strokes, heart attacks and, in the worst case, death. The gas can leak from any type of fire and even poorly ventilated household appliances such as tumble dryers can start producing carbon monoxide…the damage done to vital organs is irreversible. [173]'

10. Firearms

I once shot someone.

I was staying with a friend called Angus Rogers for half-term one year. We were messing about with an air rifle. I looked at him, raised the gun and, in true John Wayne stylee said, 'This is it, Rogers', and shot him. In the chest. The gun was empty, but what I didn't know was that a piece of mud had lodged and dried in the barrel. It didn't cut him in half or send him crashing through the wall to fall floppily dead upon the other side among the weeds, but it stung.

I'm sure I apologised, once I had stopped laughing.

This, however, is how accidents happen.

In England we thankfully have some distance from the enormous and casual use of guns that occur in other parts of the

---

[172] www.co-gassafety.co.uk. Based in Seaview, Isle of Wight, one of favourite places.

[173] www.dailymail.co.uk. 12 February 2011

world. It could be that we are simply more sensible (though I doubt it) or it could be that legislation and enforcement of that legislation is better, tighter, more effective. I also think that the majority of *sensible* Brits tend to be less dribbling with testosterone than such places as the USA or Russia and its extremely unstable satellites. I wouldn't like to put that to the test though. Frankly, there are times when I would have used a gun had I had one to hand. God help us if the wife ever gets hold of one - Barnsley will be a sea of blood that even Moses could not part.

Let's look at America first though, the home of the domesticated gun.

> 'According to the Congressional Research Service, there are roughly twice as many guns per capita in the United States as there were in 1968: more than 300 million guns in all.[174]'

As a rough estimate, that works out as one for every man, woman and child (wrong again, Chris. See below). Eleven million guns were made in 2013 alone and each school shooting or terrorist event urges people to buy more. According to Christopher Ingraham in *The Washington Post* in October 2015, for every 100 people in the USA, there are 112 guns (that's 1.12 per person, to correct my blithering self). In the UK there are 6.6 guns per 100 people. Presumably we need those for those unexpected cowboy-style call-outs that we are so prone to after revels or to guard against spiders (which *are* getting bigger every year. I saw one wearing a saddle the other evening) and bulb-mad moths. Switzerland, *Switzerland*, had 45.7 guns per 100 people. Why? What the hell is happening in Switzerland? What the hell makes a Swiss person so angry that they need a gun? Too many cowbells? Mountain-madness? The insanity of beautiful views? Geneva? Only three people live in Switzerland and one of them is Roger Moore, who I don't think has bad intentions towards anyone. www.elist10.com says that Switzerland is the safest place in the

[174] Guns In America, By The Numbers. Scott Horsley 2010. January 5, 2016

world to live.

The *New Yorkers Against Gun Violence: Fact Sheet* tends, however to put the facts and figures into a much starker light.

- 'Accidental Shootings: Guns in the home increase risk…guns in the home are 22 times more likely to be involved in accidental shootings, homicides, or suicide attempts. For every one time a gun in the home was used in a self-defense or legally justifiable shooting, there were 4 unintentional shootings, 7 criminal assaults or homicides, and 11 attempted or completed suicides.
- Deaths: From 2005-2010, almost 3,800 people in the U.S. died from unintentional shootings.
- Injuries: In 2010, unintentional firearm shootings caused the deaths of 606 people.
- More Guns = More Accidental Shootings: People of all age groups are significantly more likely to die from unintentional firearm injuries when they live in states with more guns. On average, states with the highest gun ownership levels had 9 times the rate of unintentional firearms deaths compared to states with the lowest gun ownership levels.
- A federal government study of unintentional shootings found that 8% of such shooting deaths resulted from shots fired by children under the age of six.
- Youth and Accidental Shootings: Over 1,300 victims of unintentional shootings for the period 2005–2010 were under 25 years of age.

- For kids ages 5 to 14, the mortality rate is 14 times higher in high gun ownership states than low gun ownership states.
- The majority of people killed in firearm accidents are under age 24.
- Safe Storage of Firearms: 33% of U.S. households contain a gun, and half of gun-owning households don't lock up their guns.
- Preventing Accidental Shootings: The U.S. General Accounting Office has estimated that 31% of unintentional deaths caused by firearms might be prevented by the addition…[of] a child-proof safety lock (8%) and a loading indicator [a safety device on semi-automatic handguns designed to alert the operator in some way that a round is in the chamber] (23%). The best way to avoid unintentional shootings, particularly those involving children, is to not keep a gun in the home.'

I like that last piece of advice. It smacks of the common sense that people used to have and have since lost in the headiness of McDonald's-induced chemical madness, ASDA overnight shopping lunacy and the desire to litter on even the highest mountain peaks. If you don't want your kids to shoot themselves or anyone else, don't keep a gun in the home. Yep. So unusual, it needs mentioning.

I can remember seeing an almost tearful President Obama (I resisted the urge to think that someone was digging their nails into his buttocks just off camera in order to produce the requisite tears) after the last school shootings swearing that America was going to change, it had to change. It didn't. In his time as President, there have been 162 mass murders, 18 incidents with 8 or more deaths (2005-2015)[175]. That's more even than that dolt Regan or the Dumb W. On July 24, 2015, www.nowtheendbegins.com, a site dedicated to all happy-go-lucky type things such as the end of the world, said that there had been 23 mass shootings, up to the Lafayette movie house shootings on July 23. The Orlando nightclub shooting of June 12, 2016, according to USA today, made it up to 14 mass shootings. The Atlantic (theatlantic.com) said that the 20th mass killing took place on Monday 16 September, 2013. They even created an animated map to show us what the toll looked like. That is quite a discrepancy which suggests that America, not the brightest of countries, has actually lost count of the number of killing sprees in which its kinfolk participate. In fairness, some mass killings are described as '4 deaths and above by a single shooter'. www.motherjones.com, by that criteria, judges there to have been 31 sprees. All the same…

Why do they call them 'sprees'? I thought you went on a 'spending spree', a shopping spree', or a 'drinking spree'. I thought a spree was a happy thing. Who decided that it was acceptable to go on a 'shooting spree'? The word 'spree' comes from Irish 'spraoi', meaning 'fun' or 'sport', which in turn is of North Germanic origin. It's difficult to find the moment when the words 'killing' and 'spree' were joined in this unholy matrimony. 'Killing Craze' is fine, as are 'Murderous Mayhem', 'Homicidal Hysteria' and 'Manslaughter Mischief', but to link such a happy, lamby, vernal word with, let's be honest, a fairly dark subject, seems a bit off.

For want of anything else, I'll blame the media. It's usually their fault.

I digress. This is about accidental shootings. I'm saving murders as a special treat for those brave enough to stick with this book for that long.

www.gunviolencearchive.org says that there has been a rather sobering 1,536 accidental shootings in 2016 so far (it doesn't

---

[175] www.snopes.com. June 12, 2016. A fact also shared by TruthStreamMedia.com on Dec. 2, 2015. Who nicked from who?

state how many of them were deaths. It listed 62 pages of the individuals to whom this had happened. I had counted 53 fatalities by page 9 and then began to run out of the will to live). In a truly Tarantinoesque Pulp Fiction-type incident, one of the deaths involved the accidental shooting of a lad by his friend from the back seat of a car. *Everytown for Gun Safety* has found at least 77 instances this year in which a child younger than 18 has accidentally shot someone[176]. The same article lists these tragedies:

- On April 20, a 2-year-old boy in Indiana found the gun his mother left in her purse on the kitchen counter and fatally shot himself.
- The next day in Kansas City, Mo., a 1-year-old girl evidently shot and killed herself with her father's gun while he was sleeping.
- On April 22, a 3-year-old in Natchitoches, La., fatally shot himself after getting hold of a gun.
- On April 26, a 3-year-old boy in Dallas, Ga., fatally shot himself in the chest with a gun he found at home.
- On April 27, the Milwaukee toddler fatally shot his mother in the car.
- That same day, a 3-year-old boy in Grout Township, Mich., shot himself in the arm with a gun he found at home. He is expected to survive.
- On April 29, a 3-year-old girl shot herself in the arm after grabbing a gun in a parked car in Augusta, Ga. She is also expected to survive.

Let me just repeat this: 'The best way to avoid

176 www.washingtonpost.com. May 1 2016.

unintentional shootings, particularly those involving children, is to not keep a gun in the home.'

Thankfully, it's a different story in the UK, but I do wonder for how much longer this will be the case.

> 'Guiding gun control legislation in the United Kingdom includes:
>
> - the Firearms Act 1968
> - the Export Control Act 2002
> - the Export Control Order 2008
> - the Import, Export and Customs Powers (Defence) Act 1939
> - the Import of Goods (Control) Order 1954
> - the European Council Directive of 18 June 1991 on Control of the Acquisition and Possession of Weapons, the European Union Firearms Regulation of 2012, the Common Position on the Control of Arms Brokering of 2003
> - the Convention of 1 July 1969 on Reciprocal Recognition of Proofmarks on Small Arms[177].

In America, the use of guns is controlled by a small elf-like creature called Squeem who lives in a cage underneath the dome of the Capitol Building in Washington DC and survives upon the kind donations of hundred-dollar bills from the NRA. This might not be true.

Each state in the US has its own gun laws which in itself causes difficulty in understanding of the laws (when and where can I actually carry this bazooka?) and enforcement. This is compounded by mentality. Americans believe they have the right to arm themselves according to the constitution.

This is a moot point:

> 'In United States v. Cruikshank (1876), the Supreme Court of the United States ruled that, "The right to bear arms is not granted by the Constitution; neither is it in any manner dependent upon that instrument for its existence" and limited the applicability of the Second Amendment to the federal government. In United States v. Miller (1939), the Supreme Court ruled that the federal government and the states could limit any weapon types not having a "reasonable relationship to the preservation or efficiency of a well-regulated militia.[177]'

This says that arms may be borne as part of a militia. That's it. It has nothing to do with the individual bearing arms and yet, in that random, slap-happy way that occasionally occurs within the ass's throat of the law, it has adapted to the whim of the people, despite the fact that it has been proven time and time again that a large element of those people are incapable of maintaining clean belly-buttons let alone maintaining the safety of a gun.

> 'It would be nearly 70 years before the court took up the issue again, this time in the District of Columbia v. Heller in 2008. The case centered on Dick Heller, a licensed special police office in Washington, D.C., who challenged the nation's capital's handgun ban. For the first time, the Supreme Court ruled that despite state laws, individuals who were not part of a state militia did have the right to bear arms. As part of its ruling, the court wrote, "The Second Amendment protects an individual right to possess a firearm unconnected

---

[177] United States v. Miller, 307 U.S. 174 (1939). Cornell University Law School. Retrieved September 5, 2013 and CRS Report for Congress District of Columbia v. Heller: The Supreme Court and the Second Amendment April 11, 2008

with service in a militia, and to use that arm for traditionally lawful purposes, such as self-defense within the home."[178]'

However, in 2016,

'...in Caetano v. Massachusetts (2016), the Supreme Court reiterated its earlier rulings that "the Second Amendment extends, prima facie [accepted as correct until proved otherwise], to all instruments that constitute bearable arms, even those that were not in existence at the time of the founding" and that its protection is not limited to "only those weapons useful in warfare".[179]'

They are like bees who have lost their way back to the hive and just don't know which way to turn - 'Oh, never mind,' said Mr Bee. 'I'll just go whichever way the wind blows.'

As a way to underline the American (and I'm sorry to pick on our 'special friends', but you do kind of invite comment) mentality, have a read of this from *The Guardian* of January 2013.

'Colorado stands out

In 2012 in the post-Virginia Tech era, a Colorado Supreme Court ruling allowed students and faculty to carry concealed guns on campus by overturning the gun ban at Colorado University. The concealed carry law permits carrying a gun in public except for K-12 schools but does not explicitly include college campuses.

---

[178] www.livescience.com. The Second Amendment & the Right to Bear Arms Chad Brooks. January 22, 2013

[179] Liptak, Adam (March 21, 2016). "Supreme Court Declines to Hear Challenge to Colorado's Marijuana Laws". The New York Times. Retrieved March 21, 2016.

Background check "loopholes" exist

Many states require background checks as part of the permit or licensing process, but a person could become ineligible between the time the state issues the permit or license and the sale of the firearm. Furthermore, some states do not have a formal process of revoking licenses or permits once a person becomes ineligible. Also, transactions between private sellers have fewer regulations and make it possible in some states for a sale to occur without the buyer having to undergo a background check.

Most states do not require reporting missing firearms

Most states do not require the reporting of lost or stolen firearms. In 2010, a law was passed in Nebraska that only suggests that permit holders notify local authorities if a firearm goes missing, but not under law.

Concealed carry laws are where states are the most alike

There are three tiers of regulation that allow public citizens to carry a concealed handgun. The most frequent case, a shall-issue state, will issue a permit to anyone who applies for a permit, meets a general set of requirements and passes a background check. The less frequent case, a may-issue state, extends the same requirements but allows local authorities to use discretion when granting a permit. Lastly, a select few states do not require a permit at all to carry a concealed handgun in public.

Many gun laws are subject to local discretion

Laws in many states reiterate the federal law, and a few have added subtle layers of regulation affecting some aspects of the second amendment; however, the final say is often handed down to the local level. In more than 20 states, college administrators make the final decision whether to prohibit firearms, and in other instances venues must most explicitly if guns are banned.

Gun regulation can be incredibly specific

The attention to detail in gun regulation is both incredibly specific while at the same time creating regulation that isn't fully comprehensive. Several states in the northwest, mid-west and south-east generally prohibit firearms where alcohol is served, but then include caveats that allow firearms when a specific percentage of income of the establishment goes to business not related to alcohol.

Some states prohibit registration of firearms

Laws in states including Idaho and Alaska prohibit authorities from registering firearms or enforcing any local ordinance that regulates the registration of firearms. Advocates of gun regulation say that such registration when combined with owner licensing or permitting provides the strongest means to track the possession and ownership of firearms.'

(The Great) Bill Bryson gets the last word on this. I like to quote him. I get wisdom by proxy:

'I noted that Congress had passed a law prohibiting the US Department of Health and Human Services

> from funding research that might lead, directly or indirectly, to the introduction of gun controls…The government of the United States refuses to let academics use federal money to study gun violence if there is a chance that they might find a way of reducing the violence. It isn't possible to be more stupid than that.'[180]

It is all just a recipe for disaster.

Talking of recipes, I'm getting down from my high-horse and going to have some food now.

11. Suffocation

You would think that accidental death by suffocation would be difficult to achieve. It conjures images of men being found prostrate on the living room floor after some sort of inadvertent collision with a stray pillow.

However, think of the ways. Choking is essentially suffocation, the awful Sudden Infant Death Syndrome (SIDS) is another, (auto)erotic asphyxiation can lay claim to this section, as does solvent abuse.

SIDS really only came to our attention after the very public death of TV presenter Anne Diamond's baby in 1991. The public lust for gossip and the redtops' delight in supplying it inadvertently lead to the lives of thousands of babies being saved by Ms Diamond's subsequent campaigning.

She was, to be honest, a bit Marmite. Wives hated her. Husbands wished that she could be, if not their wife, then their concubine. She took morning TV from a rather staid and amateurish idea to the commercial, if slightly tacky, success that it is now. Her public profile, and the invasion of her privacy that came with it, guaranteed that she would not be able to mourn the death of her son in privacy. However, from such frankly revolting feeding from the bones came her successful campaign to make the

---

[180] Bill Bryson. *The Road to Little Dribbling.* 2015.

country aware of SIDS. Career-wise it was a risk. No one likes to see those they admire in the rude flesh, so to speak. Once elevated, no trace of humanity may be allowed to peek through. Anne Diamond pushed this aside and charged headlong into her campaign and took the country with her. She put a very public face to an issue that had been hushed up by shame and the fear of accusation. It was easier to accuse parents of murder or neglect than for the social services to admit that there was a syndrome occurring, across the world, for which they couldn't account.

The CDC defines SIDS as:

> '…the sudden death of an infant less than 1 year of age that cannot be explained after a thorough investigation is conducted, including a complete autopsy, examination of the death scene, and a review of the clinical history.'

Similarly, the Lullaby Trust defines the syndrome thus:

> 'Sudden Infant Death Syndrome (SIDS) is the sudden and unexplained death of an infant where no cause is found after detailed post mortem.'

You will note a certain vagueness in the description. This comes from the fact that there is no cause pinpointed; nothing and no one to blame. That is not something that, as a nation and as a species, we are comfortable with. We have always managed to provide answers for ourselves – look at plague and cholera, Ebola and HIV – and when we can't do this, the instinct is to either do an ostrich, which Ms Diamond thankfully rendered us unable to do, or find a sacrificial goat to shove out into the desert of ignorance.

About 3,500 infants in the US come under the umbrella of SIDS each year[181]. In the UK it is 'around 250 babies and

---

[181] www.cdc.gov/sids/aboutsuidandsids.htm

toddlers still die every year of SIDS.[182]' In 2014 there were still 212 unexplained infant deaths in England and Wales although, that statistic alone shows that the figures relating to this particularly cruel death are lowering. That accounts for about 8% of all infant deaths for 2014. 55% of them were in boys. There is a theory called the X-linkage hypothesis which suggests that, at a genetic level, where the innate protection against cerebral anoxia (lack of oxygen to the brain) prevents death, the opposite genetic variant (the *allele*) is more prolific in males resulting in a higher death rate from this anoxia.

SIDS IN THE UK (BABIES AGED BIRTH TO ONE YEAR)				
	England & Wales	**Scotland**	**N. Ireland**	**UK Total**
2013	249	10	10	269
2012	234	29	2	265
2011	247	29	5	281
2010	254	26	7	287
2009	283	24	13	320
2008	287	22	9	318
2007	273	31	10	314
2006	285	29	11	325
2005	325	20	10	355
2004	317	31	17	365
2003	315	44	5	364
2002	298	34	4	336
2001	330	35	12	377
2000	334	35	5	374

There is no doubt whatsoever that the reason the number of deaths due to SIDS has fallen is because of education. People now know that certain types of sleeping positions, clothing, mattresses and blankets have a detrimental effect upon the

---

[182] www.lullabytrust.org.uk/statistics

vulnerable infant.

The NHS offers this advice:

- Place your baby on their back to sleep, in a cot in the same room as you for the first six months.
- Don't smoke during pregnancy or breastfeeding and don't let anyone smoke in the same room as your baby.
- Don't share a bed with your baby if you've been drinking alcohol, if you take drugs or you're a smoker.
- Never sleep with your baby on a sofa or armchair.
- Don't let your baby get too hot or cold.
- Keep your baby's head uncovered. Their blanket should be tucked in no higher than their shoulders.
- Place your baby in the "feet to foot" position (with their feet at the end of the cot or Moses basket).

* Offer the baby a pacifier when going to sleep. Pacifiers at naptime and bedtime can reduce the risk of SIDS. Doctors think that a pacifier might allow the airway to open more or prevent the baby from falling into a deep sleep. If the baby is breastfeeding, it is best to wait until 1 month before offering a pacifier, so that it doesn't interfere with breastfeeding[183].

The CDC offers these examples as pointers:

---

[183] This one from www.nytimes.com/health/guides/disease/sudden-infant-death-syndrome/overview.html September 24, 2016

– Mechanisms that lead to accidental suffocation include:

- Suffocation by soft bedding—for example, when a pillow or waterbed mattress covers an infant's nose and mouth.
- Overlay—for example, when another person rolls on top of or against the infant while sleeping.
- Wedging or entrapment—for example, when an infant is wedged between two objects such as a mattress and wall, bed frame, or furniture.
- Strangulation—for example, when an infant's head and neck become caught between crib railings.

They also supply some interesting information about the effect of SIDS upon American Ethnicities.

- SUID death rates per 100,000 live births for American Indian/Alaska Native (190.5) and non-Hispanic black infants (171.8) were more than twice those of non-Hispanic white infants (84.4).
- SUID death rates per 100,000 live births were lowest among Hispanic infants (50.8) and Asian/Pacific Islander infants (34.7).
- SIDS comprised the largest proportion of SUID deaths for all racial/ethnic groups, ranging from 48% of SUID among AI/AN and A/PI to 54% of SUID among NHW.
- Accidental suffocation and strangulation in bed comprised the smallest proportion of SUID deaths for all racial groups, ranging from 16% of SUID among

Hispanics, to 23% of SUID among NHB.'

It's interesting to note the two giant, rare albino elephants in the room here – education and poverty associated with minorities.

A 2016 article in the *New York Times* gives a slightly more comprehensive list of potential vulnerabilities in infants:

Sleeping on the stomach
Being around cigarette smoke while in the womb or after being born
Sleeping in the same bed as their parents
Soft bedding in the crib
Multiple birth babies (being a twin, triplet, etc.)
Premature birth
Having a brother or sister who had SIDS
Mothers who smoke or use illegal drugs
Being born to a teen mother
Short time period between pregnancies
Late or no prenatal care
Living in poverty situations

They put poverty on the list and if the list is examined, there are several associations with poverty and low education – cigarette smoking, teen parenting, late or no prenatal care, the use of drugs, short time between pregnancies. Yes, you're right, they do occur in white folk with money, but less so.

> 'Pickett and colleagues examined the effectiveness of the Back-to-Sleep[184] campaign as a function of social class. Social class was measured by maternal educational achievement The researchers used data

[184] The Safe to Sleep campaign, (formerly the Back to Sleep campaign), is an initiative backed by the US National Institute of Child Health and Human Development (NICHD) at the US National Institutes of Health to encourage parents to have their infants sleep on their backs to reduce the risk of sudden infant death syndrome.

sets of all infant deaths caused by SIDS for the years 1989-1991 and 1996-1998 from the U.S. National Center for Health Statistics linked birth and death certificate data on all infants born in the United States and those who died in the first year. In sum, the study found that social class inequities in SIDS, measured by maternal education, did not narrow after the Back-to-Sleep campaign compared with the pre-campaign era. Absolute risk of SIDS was reduced for all social class groups, though a widening social class inequity was evident as women with more education experienced a greater decline than women with less education.
...the educational efforts were expanded to include additional community organizations so as to achieve greater penetration into the black community...This effort will increase knowledge; an important facet of a successful intervention.
A recent study in the United Kingdom analyzed five years of case control data to highlight the changing etiology of SIDS. The results indicated that the Back-to-Sleep campaign had been successful in the United Kingdom in reducing the rate of SIDS. However the epidemiology of SIDS has shifted...with SIDS occurring increasingly from low SES [socioeconomic status] families that also included an increase in single parents, younger mothers, and low birthweight infants.'

Child Poverty in America Today - By Barbara A. Arrighi, David J. Maume

So, once again, the ugly dragon of low education and low income rears its head and burns those without.

To change the subject a little, let's look at death-sex or autoerotic asphyxiation as the professionals would have us call it.

This was a phase that the world seemed to be going

through just a few short years ago. Those who were famous and succumbed to this were Michael Hutchence, the lead singer of INXS and David Carradine, cool movie dude. One of them was found hanging on the back of his door, the other in the wardrobe. There was also Albert Dekker (actor), Hideto Matsumoto/Hide (Japanese singer), Stephen Milligan (politician and journalist), Kevin Gilbert (musician), Frantisek Kotzwara (18th century composer), Reverend Gary Aldridge (Baptist Church Minister – hogtied and in two wetsuits – '…face mask which has a single vent for breathing, a rubberized head mask having an opening for the mouth and eyes, a second rubberized suit with suspenders, rubberized male underwear, hands and feet have diving gloves and slippers'[185]) and Vaughn Bodé (Graphic Artist).

Why?

Seriously.

Why?

Well, according to *Psychology Today* (Feb 27 2014 – so not quite today):

> 'Erotic asphyxiophilia connects two sources of euphoria: orgasm and hypoxia (lack of oxygen). Swimmers, who have nearly drowned, say they felt elation before losing consciousness; this is hypoxic euphoria.'

Essentially, the chemical reactions within the brain and the body have a similar effect to an opioid euphoria.

Kinky high. There you go.

I'm off for a deep bath.

As the water runs, I'll wind this section up with some stuff about choking, to which we are all surprisingly vulnerable.

The physical process of choking is really just where something blocks the airway, as shown in the following diagram.

---

[185] www.therichest.com/rich-list/most-shocking/10-well-known-men-who-allegedly-died-from-erotic-asphyxiation/

If the airway is compromised to the extent that the oxygen intake is reduced for a prolonged period, then the lowered oxygen levels in the body will no longer be compatible with life. There is also the risk of laryngospasm, where the larynx is sent into involuntary muscular contraction.

The queen mother was particularly adept at blocking her airway. She managed it three times, requiring on two occasions a visit to theatre to have the offending obstruction removed. One has to wonder at what stage it goes from accident to hobby. George W Bush too had a famous choking incident.

In *What Do People Die Of? Mortality Rates and Data for Every Cause of Death in 2011 Visualised*, the guardian, in a wonderfully comprehensive list of deaths and their causes, availed us of our constant susceptibility to the man with the scythe. It included the inhalation of stomach contents (*au* Hendrix, I assume) and the inhalation and ingestion of food (*à la* Queen Mother) as part of its statistics, both under the choking umbrella.

Interestingly, the Hendrix Stylee had gone down by 7.56%. That's quite a lot. Startlingly, inhalation and ingestion of food had gone up by 16.67%.

Cause	Type	MALE	FEMALE	2011	2007	% change, 2007-2011
Inhalation of stomach contents	**Choking**	**59**	**54**	**110**	**119**	-7.56
Inhalation and ingestion of food	**Choking**	**100**	**103**	**210**	**180**	**16.67**

Did the old dear start a trend? Had people suddenly stopped partying and at the same time started to eat massive amounts of chicken on the bone? Had we come to expect boneless food to such an extent that the body, upon finding said artefact, was startled into a death-choke-spasm? Are we all suddenly eating too quickly and talking at the same time? Has the decline in simple table-manners led to an increased death-rate?

I don't know. All I know is that I will advise my wife to be more careful about her choice of food with a view to sustaining life. For my part, I will chew more carefully.

The bath is full. I'm off. Might see you later. Might not.

12. Fires

> 'For women childbirth remained the greatest danger to their lives; yet, barring complications, the second leading cause of death among women remained death from fire. Open dairies and hot stovetops were everywhere, and long skirts made of flammable materials, and children's clothing of cotton or linen, were easy targets for accidental injury and death. Many more people lost their lives to fire than to Indian attacks on the frontier. Moreover, with buildings made largely of wood, towns fires were almost inevitable. Fire destroyed dozens of buildings annually even in small towns, and a single devastating incident could destroy whole communities. Virtually the entire town of Tombstone, Arizona, burned three times in as many decades.[186]'

Accidental fire has always been high on the list as a cause of death. When we think of the great fires of history – London, San Francisco, Rome, Constantinople, Boston, New York – it's a wonder really that the civilised world hasn't simply been consumed by our carelessness.

The risk of death by fire to the individual (and by way of subsequent twirling and falling into further combustible materials, the population) has always been a consequence of simply waking up and doing the daily routine.

> 'Clothing catching fire was the most common cause

---

[186] *Family Life in 19th-century America.* James M. Volo, Dorothy Denneen Volo

of accidental death for toddlers and young children, and occurred mainly in this age group. The following case is typical. In 1816, Martha Draper, aged 3, was alone by the fire in her parent's kitchen and the linen apron and petticoat she was wearing caught fire. Boys also died in this way. In 1801, Thomas Stangroom, aged 4 and a half, was left alone in his mother's house in St Paul, while she went to the baker's for bread, and he set fire to the linen slop he was wearing.[187]'

Over time, legislation was passed by kings and governments to prevent fire. Even William the Conqueror had a hand in it by declaring that all fires be extinguished at night. A metal cover, a *couvert feu*, was used. This became the word 'curfew'[188].

In 2014/2015 there were 263 fire fatalities in the UK. There were nearly 8,000 injuries from fire. 41% of these fatalities were in the over 65 (49% when only in the home), which is a glaring pronouncement that the older you get a) the slower you are to move b) the more likely you are to fall asleep with a fag in your hand and c) when you sleep, boy do you sleep. 40% of the deaths were actually from inhalation of gas or smoke, 33% from burns. With that little fact comes a horrifying realisation – the victims who died of inhalation probably died in their sleep, unconscious and, hopefully, unaware. Those that burned, and it makes me shudder, were possibly more aware of their fate. That is cruel, to say the least.

As for the causes, well, I wasn't kidding about the fags. 36% of accidental fires in the home were caused by smokers' materials in 2014/15. That is over a third. That's madness. Cooking appliances caused 50% of the blazes[189]. It's easy to see how the figures in the elderly can be so high when one takes into

---

[187] *Accidental Death In Nineteenth Century Norwich* - Jill Waterson, 2008

[188] http://www.fire.org.uk/history-of-fire-safety.html

[189] Fire Statistics England, 2014/15 29 June 2016 Home Office

account dementia, slower reactions and debilitating illness.

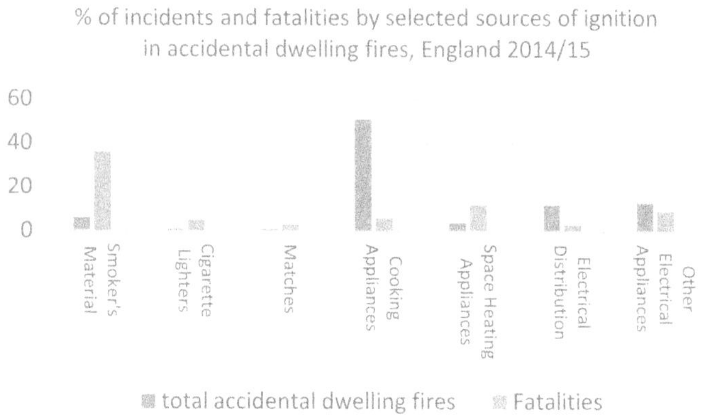

These are UK figures. In 2013, America had 1,240,000 fires which led to 3,240 deaths. The cost was $11.5 billion[190]. The good news is that these were down on previous years. 50% of residential fires in the USA were caused by cooking, 12.5% by heating appliances and 6.3% by electrical malfunction. Smoking caused 2%.

With the massive rise in the production of electrical equipment over recent years, it's not surprising that electrical faults as a cause have become higher.

> 'Modern living has meant we use more and more electrical appliances in the home. For instance, just 20 years ago the average UK home had a hi-fi system and one TV or video, whereas today it is more likely that there are at least two TVs, a DVD player, a satellite receiver, games console, microwave and computer. So the risk of electrical accidents in the home is much higher than before… Government statistics show that

[190] https://www.usfa.fema.gov/data/statistics/

> electricity causes more than 20,000 fires a year - almost half of all accidental UK house fires. Each year, about 70 people are killed and 350,000 are seriously injured due to an electrical accident in the home.[191]'

As a species, we do seem to have a knack for finding new and exciting ways to kill ourselves. The London Fire Brigade says that in 90% of cases, a fire was caused by a fault involving 'white goods' - dishwashers, washing machines, tumble dryers, fridges and freezers. When we think about how many more things we are plugging in nowadays – game consoles, TVs, Sky boxes, digital recorders, Freeview boxes, printers, computers, powerful vacuum cleaners, drills (my father had a *hand drill*), food mixers with the power of helicopters, all those things we leave on stand-by, even bicycles and cars – and yet not updating our homes to cope with the needs of these powerful beasts, it's not really surprising that we are leaving ourselves vulnerable.

13. Poisons

The Telegraph

Warning over mushroom foraging after more than 80 people are poisoned

Public Health England warned more people could fall ill as the foraging season begins in earnest.

By Alice Philipson
02 August 2014

---

191 www.electricalsafetyfirst.org.uk/guides-and-advice/around-the-home/how-safe-is-your-home/

> **Foragers are being warned to watch out for toxic mushrooms after dozens of people fell ill from poisoning. Public Health England (PHE) said more than 80 cases of mushroom poisoning have been reported so far this year with the foraging season only just under way... Nicholas Evans and three family members had to be put on kidney dialysis after eating toxic mushrooms they had gathered on a woodland walk. Across Britain, foragers have been spurred on by celebrity chefs like Hugh Fearnley-Whittingstall and Antonio Carluccio, who have encouraged people to pick field mushrooms and other delicacies. As the trend for organic, home-grown food has caught on, many people have started hunting in their local woods for edible varieties. However, last year, PHE's National Poisons Information Service (NPIS) recorded 237 cases of poisoning across the UK, with many involving children under the age of 10. Just one month into the 2014 foraging season, which typically begins in September, the NPIS has been consulted for advice on 84 cases.'**

In my opinion, anyone who follows Huge Farting-Witless-Fool's constantly pompous advice is asking for trouble. I trust the Carluccio feller a bit more, but having seen him on Saturday Kitchen, I realise that he will eat pretty much anything put in front of him, feathers and all.

I don't like eating out. This, I realise, makes me poor company. In my defence, I will say that I have had two bouts of serious food poisoning in my life that have left me mighty paranoid about where I eat. The first dose was from a hospital kitchen – the sausages were dodgy. I had salmonella and lost three stone and three weeks of my life. The second one was from a

Wendy's restaurant in London. I can't see a dirty fork anymore without breaking out into a cold sweat.

My daughter worked at a very popular hotel/restaurant in Barnsley, **not far from a golf course**, and described the conditions in the kitchen there to me. It merely added fuel to my already hot, hot fire. I also knew a chef who told me some absolute horror stories including pissing in the soup, spitting on the food of those who complained and retrieving spilled food from the floor. One man I worked for, who repackaged frozen food, actually picked some potatoes out of a bin and sold them on as part of a ready meal.

It amazes me that we can trust strangers to make food for us and amazes me still more that we pay them to do it. Anyone who said to me, as I sat down and perused the menu, beaded by sweat and jerking with palpitations of fear, 'We picked the mushrooms ourselves this morning while the dew was still fresh on the ground', would get short shrift. 'Show me the tin whence the mushrooms came,' I would demand. 'I want none of your foraged death on my plate, sir'.

In the table below is a summary of fungally consequences[192] and I don't know about you, but words like 'non-lethal' and 'potentially deadly' hold little sway with me; anybody who goes around the countryside picking at plants to eat simply because they think they'll be better than the ones in Sainsbury is deluded.

Toxin	Toxicity	Effects
Alpha-amanitin	Deadly	Causes liver damage 1–3 days after ingestion.
Phallotoxin	Non-lethal	Causes gastrointestinal upset.
Orellanine	Deadly	Causes kidney failure within 3 weeks after ingestion.
Muscarine	Potentially Deadly	Can cause respiratory failure.
Gyromitrin	Deadly	Causes neurotoxicity,

192 https://en.m.wikipedia.org/wiki/Mushroom_poisoning

		gastrointestinal upset, and destruction of blood cells.
Coprine	Non-lethal	Causes illness when consumed with alcohol.
Ibotenic acid	Potentially Deadly	Causes neurotoxicity.
Muscimol	Non-lethal	Causes CNS depression and hallucinations.
Psilocybin and psilocin	Non-Poisonous	Causes CNS arousal and hallucinations.
Bolesatine	Non-lethal	Causes gastrointestinal irritation, vomiting, nausea.
Ergotamine	Deadly	Affects the vascular system and can lead to loss of limbs and death.

'Affects the vascular system and can lead to loss of limbs and death.' I'll have two please.

Why would you allow someone who was possibly pissed or high less than twelve hours ago and who might, for all you know, have set out on his first death-forage that very day, with a copy of Huge Farting-Witless-Fool's *Road to Fungal Death* tucked under his arm like a bible, decide what was right or wrong for you to eat? Why would you trust a twelve-year-old oik on an apprenticeship to make sure that your red kidney beans had been properly soaked overnight so that you didn't die a painful, pointless and undignified death while watching your liver descend *per rectum* towards the floor?

I realise though, that this might be my own bias weeping none too subtly through the pages. Being poisoned by strangers does this to a chap.

Mushrooms, however, are not the only fruit. There are many ways that we accidentally poison ourselves. You would think that it wouldn't be, what with all the hermetically sealed foods and beverages we have today and all the warnings on the packets. On the other hand, many people still smoke and look at those disgusting pictures they put on fag packets.

Now bleach comes in a bottle that, to any normal person is in fact unopenable. You have to have as many fingers as a spider has legs and the prestidigitation skills of David Blaine to get into one of these and yet, in the United States, household bleach is the number one cause of accidental poisonings, with more than 50,000 cases (including eight deaths) reported to poison control centres in a single year[193]. How? How do you accidentally ingest bleach? How many people stand at the sink, emit a deep, tired sigh and say, 'Oh, not again! I've drunk bleach.'?

Less than two, I bet.

According to WHO[194], in 2004 acute poisoning caused more than 45,000 deaths in children and youth under 20 years of age – 13% of all fatal accidental poisonings worldwide. That is a large amount. I must be the only one who struggles with the lids. Fatal poisoning is highest (strange, considering child-proof lids) in under one-year olds and it is four times greater in low-income countries than high and middle-income countries.

Obviously, we're not just talking about bleach here, we include carelessly discarded medication, household products, pesticides, poisonous plants and bites from insects and animals.

In the UK in 2002, almost 31,500 children aged under 15 went to hospital after suspected poisoning – over 26,000 of these were under five years old. About 7,000 children under 15 were admitted to hospital. 69% of the admissions were associated with medicines – a quarter of these being commonly used drugs such as analgesics[195]. Other things often ingested by children include aftershave and perfume, hair remover, nail polish and nail polish remover, soap (who hasn't tasted Matey bubbles?), washing up liquid, fabric conditioner, clothes/dish washer tablets and liquids, paint thinner, paraffin and furniture and floor polish[196].

It is interesting to note that, once again boys have higher rates of poisoning than girls and once again it can be put down to

---

193 *Bleached to Death (Again).* Deborah Blum. http://blogs.plos.org

194 *Children and Poisoning* – WHO and UNICEF

195 Child Accident Prevention Trust factsheet.

196 Scientific Committee on Consumer Safety (2009).

the idiocy of lads who want to impress their peers by taking risks and perhaps because, in low-income countries, where Health and Safety is not a priority and income is, they are forced, at a young age, to take more risks in the workplace.

Why children under one year? I have three children and during the first year of their life everything that came into their hands was food and went straight to their mouths. Children of that age are innately curious and have absolutely no concept of actions and consequences.

Another obvious cause is the storage of such substances. How many parents out there have spent hours with drips of sweat running down their faces as they contorted themselves into ridiculous positions putting cupboard locks on every cupboard in the kitchen? How many parents have found out that their kids have managed, like some ninja burglar, to find their way through these within twenty-four hours?

We in the west are very careful about how we store these things and are even regulated by law. We are privileged in that respect. In low-income countries, once again, it is a matter of priorities - it's about the quickest way of making money in the shortest time possible and the risks are very often not taken into account.

14. Motor Vehicle Accidents

On my first day as a bus driver, at about 0550, I reversed into a very sturdy pole on a road sign in Darfield, an ex-mining village just outside town. I was so nervous, I had missed my turn and had to back up. A large amount of the rear end of the bus had gone missing. The passengers got to travel free too, because I forgotten to log in at the depot. It was a really shitty start to a job.

I went back to the main office at the bus station and reported my 'incident' to Al, a very philosophical and easy going lead driver who took no shit from anyone, because he didn't want to. He came with me to where I had parked up and stood at the rear of the bus.

'Fookin' 'ell!' he said. 'Not to worry, we'll soon knock that aht. Are you reyt to drive?' Bless his kindness.

A couple of days later, I missed another turn and felt compelled to reverse again, this time into a residential side-street. I near beshit myself, as they would say in the eighteenth century.

I would have been better off just driving in reverse, to be honest. I must have filled out more accident forms than Winnie-the-Poo has had jars of delicious honey. I think the guys in the office must have done some covering for me; many people got fired for less. Thanks chaps.

I was a dreadful bus driver. I don't think there was a day went by when I didn't somehow beshit myself due to lack of concentration or poor memory. I remember one night, in a place called Penistone, where we had to take the bus through some very windy back streets I, you won't be surprised to know, missed my turn. I only realised this important fact once I had come to the end of the cul-de-sac.

I had weaved through many cars to get into this predicament and there was no way I was going to get out of this place facing the right direction. I contacted the office and told them what I had done. 'What the fook does tha want me to do abaht it? Tha got thissen in there, get thissen aht.'

You know that hot and cold feeling you get when you are certain that you are going to die, even though you aren't? I had that. I sweated and froze at the same time. Palpitations fell from my mouth like rotten teeth.

I reversed that bus through badly parked vehicles and twitching curtains and made not one error. It was a fucking masterful piece of driving.

I think by this stage I was beginning to realise that I was in the wrong job.

What is the point of telling this ditty?

Only to show you how easy it is to have an accident. I ran someone over. I was at some lights at an intersection, the lights changed and off I went. I had my eyes on some kids who were on bikes near the road, not on the silly cow who decided to cross the road against the lights while looking at her phone. Boy, did she fly.

I thought she was dead. She made that crushed melon sound against the side of the bus, spun 270°, then went down like a drugged whippet.

She survived.

You see? It's so easy to be looking at what you think matters when there's an idiot a few feet away who thinks that Gala Bingo is more important than road safety.

I saw a lorry driver on the Huddersfield Road reading while he was driving without his hands on the wheel. I have seen women doing their makeup and men shaving while they drive. I have almost knocked down I don't know how many Barnsley College students because they have the road-sense of the aforementioned whippet.

In 2011, 5,419 people were killed or injured in vehicle accidents (most of them students on their way to Barnsley College, no doubt). Of those, 2,776 died. Motor accidents are the biggest cause of death in 15-24 year olds and 40% of males have an accident within the first six months of driving. I did. Mind you, I never stopped having accidents. An 18-year-old is three times more likely to have an accident than a 48-year-old[197].

A lot of this is down to inexperience and showing off. Any males[198] reading this cannot deny it. We all think that we are the best driver that ever got behind the wheel of a car and that stunt driving is actually part of the evolutionary process as natural as producing insulin. They tend to forget about the diabetics out there. And just for those who turn their nose up at me questioning their masculinity and the pool of testosterone pulsing through them, the RAC says:

> 'There is also a significant and notable disparity between the deaths caused by road traffic accidents between men and women, with men being over three times as likely to die from a road accident.'

---

[197] www.itv.com/news/update/2012-11-19/road-accident-single-biggest-cause-of-accidental-death-of-young-people/

[198] Image: RAC Foundation: *Mortality statistics and road traffic accidents in the UK*

Corresponding numbers of deaths in the UK from road traffic accidents, 2009			
AGE	**MALE**	**FEMALE**	**ALL**
0-4	10	11	21
5-9	17	6	23
10-14	29	19	48
15-19	259	68	327
20-24	273	59	332
25-29	207	34	241

In 2012, the Daily Express declared:

'It's official...women are safer drivers than men

MEN may not like reading this but women are officially better behind the wheel[199].

They break speed limits less, brake less suddenly and avoid danger after dark by driving considerably less often at night than male motorists....experts found women drive 28 per cent less at night than men, exceed speed limits 12 per cent less and brake hard 11 per cent less...The verdict comes from the study of data from black box recorders in the vehicles of 19,000 motorists which were used to monitor more than 40 million journeys and 154 million miles of travel over four years. Half the drivers were men and half were women.

Just to level the playing field, I will repeat this from moneysupermarket.com in 2014:

[199] http://www.express.co.uk/news/uk/356997/It-s-official-women-are-safer-drivers-than-men. Nov 9, 2012

'"Speculating about the relative driving performance of men and women can be a minefield, but the traditional explanation for the discrepancy in the size of claims on gender lines is that men drive more often on motorways and fast trunk roads for work purposes, so if they have a crash it will be in busy traffic, at high speed, and will result in greater damage and injury. The corollary is that more women than men have low impact bumps and scrapes in car-parks. Additionally, it could be the case that men drive more expensive cars.'

But the Daily mail went and said this:

'It's official! Women drivers ARE more dangerous behind the wheel, scientists discover

Researchers looked at 6.5million car crashes that took place in the U.S. between 1998 and 2007

The results are even more surprising as men spend more time driving than women[200].'

Oh, dear. There's your truth, my truth and *the* truth. It is always a point of contention and interest as to which of the sexes is the better driver, but all I can say is, most of the female bus drivers I worked with were *way* better than me.

---

[200] Graham Smith for MailOnline. 7 July 2011

In 2015 there were 22,137 people reported **seriously injured** by RTAs (Road Traffic Accidents), with a total of 186,209 casualties. Up to March 2016, from March 2015, there had been 1,780 road deaths. The figures and stats don't point out any single cause for these accidents, but there are found to be a number of factors that influence RTAs:

- The distance people travel (which is partly affected by economic externalities)
- The mix of transport modes used
- Behaviour of drivers, riders and pedestrians
- Mix of groups of people using the road (e.g. changes in the number of newly qualified or older drivers)
- External effects such as the weather, which can influence behaviour (for instance, encouraging / discouraging travel, or closing roads) or change the risk on the roads (by making the road surface more slippery). I'll vouch for this. My bus skidded on ice and went straight into a covered bus stop. Bus stop? Fine. Bus? Substantial amount of front end gone[201].

The figures have been steadily going down over the years, however. In 1970 there were 7,449 deaths on the roads. In 1980, there were 5,953 and in 1990 there were 5, 217. In 2010 there were 1,857 and, as stated, in 2014, 1,775[202]. Interestingly, the only road user group to increase its death rate were, as my wife, who works on a renal unit calls them, 'donors' - motorcycle riders. Why? Motorbikes are now more powerful than they have ever been, they are more popular, especially scooters among the lads, and the

---

[201] Department for Transport: *Reported Road Casualties in Great Britain: Main Results 2015* And RAC Foundation: *Mortality statistics and road traffic accidents in the UK*
[202] *Reported road casualties in Great Britain*: Various years. - Publications - GOV.UK. www.gov.uk.

campaigns to watch out for bikers, such as 'Think Once. Think Twice. Think Bike', are less prevalent today. The number of car owners has gone up, which might account for it. Once again, we are looking predominantly at males who really have no idea how to separate their handlebars from their willies.

I constantly rage about cyclists, the way they ride two abreast, skip traffic lights and undertake. Lorry drivers have taken a large share of the blame for the stupidity of cyclists. I firmly believe that all bikes should be registered, more severe fines imposed for any infringement of the laws of the road and a road tax charged. The problem is, that isn't politically correct or going with the current political agenda. I would also ban Lycra.

The comparison between modes of death (below) in the 15-24 year old age range shows a marked difference between modes.

15-24 YEAR OLD DEATHS BY CERTAIN CAUSES 2009	No. of deaths
Assault By Firearms	7
Assault By Sharp Or Blunt Object	31
Intentional Self-Harm By Hanging	205
Accidental Drowning And Submersion	38
Smoke Fire And Flames/Heat & Hot Substances	10
Exposure To Forces Of Nature	11
Accidental Poisoning	157
Road Traffic Accidents	659
Number Of Deaths From All External Causes	1755

The lad I described at the beginning of this book, my first death, was a RTA. Some things you don't forget. Some things, it's better not to forget.

NB On the local news this morning (27 September 2016), it turns out that twice as many people are dying from drug use than from RTAs. Whether this was intended to be some sort of comfort statement before I got into my car to hit the early morning traffic, I'm not sure. Just thought I'd share.

15. Falls

> 'A fall is defined as an event which results in a person coming to rest inadvertently on the ground or floor or other lower level.[203]'

That has to be the politest, most accurate definition I have ever come across.

I'm not talking about the throwing of oneself from a tall building type of fall here. This is about ladders and chairs and tripping over kerbs, about getting blown over by strong gusts of wind and falls downstairs. Rod Hull (he of Emu fame) fell from his roof while adjusting the TV aerial.

Falls tend to matter less if you are younger; your bones bend a bit and your skin is a nice rubbery cushion, but once you have the honeycomb bones of osteoporosis and the skin of fragile, sun-baked Egyptian parchment, these things tend to hit you a bit harder.

I have looked after several people who have been blown over on the wild and windy streets of a Yorkshire winter and have sustained fractured arms and broken pelvises (pelvi?). I know of two people who have broken their necks falling downstairs and one who fell downstairs through their glass front door and nearly ended up sliced, diced and dead. I know of two instances where two individuals have managed to fall to the extent where they have managed to get foreign objects lodged in their backsides. One had an orange (they're not the only fruit, you know), somehow rectally ingested after a fall from a step ladder, the other a (wait for it) eight inch iron rasp, under pretty much the same circumstances. Step ladders have a lot to answer for. Neither of these, however, were fatal.

You might be as surprised as I was to find out that falls

---

[203] http://www.who.int/mediacentre/factsheets/fs344/en/

are the second biggest cause of accidental deaths/injury in the world. It is estimated that 424,000 people around the world die from falls a year. 80% of these are in low and middle-income countries, again suggesting a lack of resources and education. There are 37.5 million falls a year that require medical attention[203]. You would think that we would be stepping over capsized people lying on their backs like stranded woodlice. That is a ridiculous number of people. In the UK alone, every year 280,000 people end up in A&E after a fall, costing the NHS £1.5 billion[204].

The focus tends to be upon the two extremes of age; one because they are learning how to walk, the other because they have forgotten how to walk.

Children fall because they are evolving developmentally, are very curious and have a natural desire to be independent. Around 10 children die as a result of falls each year - some from windows and balconies and the remainder mostly from stairs. Falls are by far the most common causes of accidents in the home; they account for 44 per cent of all children's accidents[205].

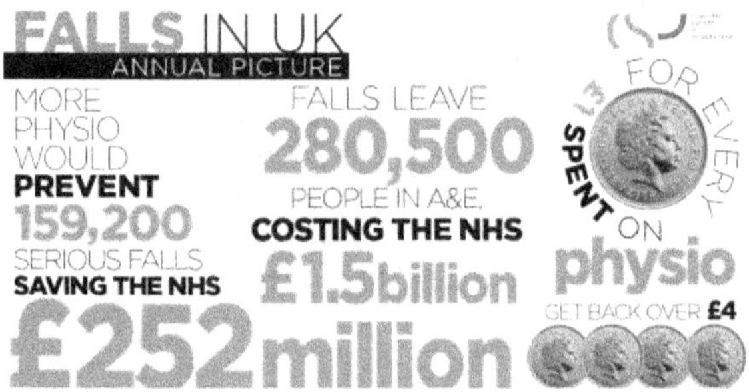

One in three people, a third, over the age of 65 fall at least once a year; at least half of those over 80 years fall at least once a year. [206]

The cost to the NHS is over £2 billion a year (£4.6

---

[204] Chartered society of Physiotherapy - *The Cost Of Falls*

[205] www.rospa.com/home-safety/advice/child-safety/accidents-to-children/#falls

[206] Image: Chartered society of Physiotherapy - The Cost Of Falls

million a day just for the over 65s), and then the hours in lost productivity for carers who have to take absence from work with the concomitant costs to industry have to be added on. The costs will go up too. The population is getting older. Hip fractures, which are the commonest consequence of falls in older people, affect about 60,000 people a year and costs about £1.7 billion, with up to 14,000 deaths as a result[207]. I have seen sprightly, mentally alert people descend into confusion, from which they never emerge, due to the psychological and physical effects of a fall. They are prone to DVT, chest infection, urine infections, constipation to the point where it can actually become life-threatening and bed sores which can lead to septicaemia - the list is endless.

And this could be just because some bugger-lugger at the council decided that that particular paving stone was safe or couldn't be arsed to lay it properly in the first place.

There are however some, shall we say, more wayward falls, which fall into a category all of their own, as I found at www.theweek.co.uk:

> 'Twelve people have died so far this year on their quest to get the perfect snap, compared to eight killed in shark attacks. The latest victim of the selfie craze was a 66-year-old Japanese tourist who died after falling down stairs while attempting to take a picture of himself at the Taj Mahal.
>
> Here are five ways that people around the world have been killed while taking a rash selfie:
> Electrocuted on top of a train
>
> A 21 year-old man was killed when he climbed onto the top of a stationary train to take a selfie with friends and accidentally touched a high-voltage power cable running overhead. The Local says that

---

[207] www.ageuk.org.uk/latest-press/archive/falls-over-65s-cost-nhs/

the Andalusian man was killed instantly by the 35,000 volt charge, but another member of the group survived the shock.

Plunge from cliff edge

A Polish couple living in Portugal with their two young children stumbled off a cliff edge while trying to take a selfie. The unnamed pair fell hundreds of feet from the scenic Cabo da Raca coastline, after apparently crossing a safety barrier in their quest for a striking picture.

Accidental gunshot to the head

***** ***** ******* made international headlines after he accidentally shot himself dead while posing for a selfie with a handgun. The 21 year-old Mexican had been drinking with friends when he decided to take a picture with the weapon, which discharged, killing him with a bullet to the head.

Fall from 30ft-high railway bridge

A Russian teenager who scaled a St Petersburg railway bridge in a fatal attempt to take a spectacular photo plummeted 30ft to her death after losing her balance. Amateur photographer ***** ********* lost her footing on the bridge and grabbed onto a nearby cable, which turned out to be live. She received an electric shock and fell to her death.

Car crash during radio singalong

A US woman died in a highway accident just seconds after uploading a selfie of herself enjoying

Pharrell Williams' hit *Happy* on her car stereo. The Independent reports that ******** *******'s status declaring her love for the tune, accompanied by a selfie taken at the wheel, appeared on Facebook just one minute before police received reports of a collision involving *******'s vehicle.'

So, that is the fifteen types of accidental deaths dealt with. We haven't mentioned the fact that champagne corks kill almost twenty-four people each year, that ants kill up to fifty people every year and that hippos kill about three thousand people a year[208].

It is really easy to be flippant about these things. I live in England and have often wondered how I was born so lucky to be so. I could have been born to a woman from a low-income country or from an underprivileged minority and been vulnerable a hundred times over to all the things that have been discussed. I am however protected by free speech, law and choice, by the rapidly withering NHS, by those who tirelessly campaign to improve our lot and those who take their time to print all these statistics so that we might know how well off we are and where we can improve.

I'm a lucky guy.

---

[208] https://likes.com/weird/the-15-most-common-accidental-deaths

PART 5

Murder

Murder is one of those things that we would all like to do (just admit it) and yet something for which we will happily condemn others. It brings out the hypocrite in us quicker than Christmas Eve brings out the big, blousy softie.

We attribute it to any one of a number of causes – love, anger, war, self-defence, a moment of madness (that lasted long enough for you to go out, buy a Bowie knife and go on a two-day hunt for the guy who cut you up on the M1 two years ago), for the good of others –

> 'About a half (48%, or 247 offences) of all homicide cases in the year ending March 2015 resulted from a quarrel, a revenge attack or a loss of temper. This proportion was higher where the principal suspect was known to the victim (59%), compared with when the suspect was unknown to the victim (33%). A further 7% (35 offences) were attributed to irrational acts and 4% of homicides (19 offences) occurred during furtherance of theft or gain.[209]'

and yet there is only one reason, as far as I am concerned; we are animals.

Now, I hear you. Really. You're saying: But animals don't kill for pleasure, only humans do that. Not quite true, but I'll go with it. However, at our most basic, in that hidden, all-powerful, genetic way, we are animals; we have just learned to be a bit more refined about it, a bit more selective, a bit shrewder. We murder for no one but ourselves, make no mistake, for that tiny string of proteins that pushes us on day by day until we shuffle off this DNA coil. There is no such thing as altruism.

If I may, in order to clarify, I would like to use a quote from a book of my own, *Condition of Life*[210]:

'It's the nature of people.

---

[209] www.ons.gov.uk

[210] Yeah, it's out there. Amazon.

I am an Animalist. I don't know if that's a thing, but it is my philosophy. I believe that humanity is only a veneer, a very thin veneer at that. Beneath that veneer, we are no more than wolves, apes, tigers or snakes (choose an animal to suit yourself should you wish). We shit, we eat, we copulate, we die. In between times, we do all we can to savage those around us, albeit in the politest way possible.

After all, what is a smile but a showing of teeth, a warning?

I've seen on the news today that it's snowing in America. Snowing. That's the fluffy bits of rain, not volcanic ash or radioactive fallout or alien attack. Snow.

People have cleared the shelves of shops in a panic buy. Why? Do they think it will never stop snowing? Are spring and summer cancelled? Regardless of the needs of others, of the elderly, of the disabled, of those who cannot get to the shop for whatever reason, they have emptied the shelves. I bet some of them have even bought aubergines (eggplants to the yanks). They would never, ever look at an aubergine but, because of the impending End of the World, they have cleared the shelves of them. Of everything. They are looking after number one. They are growling over the carcass.

Fuckwits.

In a few months, there will be a mass disposal of aubergines as people suddenly realise their stupidity. The refuse system will not be able to cope. There will be a mushy, off yellow aubergine mountain.'

Murder as a concept exists only in law. Its prevention is not an inherent, instinctive thought. If it was, we wouldn't need to make a law to try to stop it. We are now, after millennia of drip,

drip, drip brainwashing, only just beginning to recoil from the concept, though the thought goes through most of our minds on a daily basis. The concept is part of a culture that has developed over hundreds of thousands of years and was introduced to protect the wealthy weak and the financial interests and assets of those with money and power. It had nothing to do with kindness or God or Cain and Abel. Eventually, through a series of introductions into culture and law, brought about by the self-interest of the lower classes ('We will overthrow you, Lord,' said the quivering yet angry genetic mass, 'if you don't protect us'), we now all have the protection of the rich and powerful.

So we think.

Yet sometimes, the animal will out. Individuals within humanity cannot resist picking on the weak, subjugating others in order to elevate themselves and the there comes a moment when the consequences are irrelevant. We will kill because we think, in our mad, anger-fuzzed minds, for just half a second, that we will get away with it. Figures show that quite often, we do get away with it. Others, such as, at the other extreme, governments, will use murder as a part of policy. They will try to hide it because they need to maintain the status quo, they will give it another name such as 'war' or 'terrorism' and try to blame others, to leave the people to feel unthreatened, because our pretence at democracy demands that if they get caught, they will be prosecuted or, worse still, voted out.

Feet out of stirrups. Climbing down off mightily burdened high horse.

It is with this subjugation of the weaker parts of society in mind that I'm tackling this whole murder thing. It is such a vast subject. To paraphrase Douglas Adams: 'Murder is big. You just won't believe how vastly, hugely, mind-bogglingly big it is.'

There is no way that I could get everything about murder into these pages. You would have to pay so much more for this book, for one and secondly, I'm not sure I have so many years left that I can afford to dedicate the vast amount of time required to do justice (sorry) to such a subject. This is, after all, *A Beginner's Guide…*, not a round the world tour.

So, I will concentrate on:

a) Race and homicide
b) Socioeconomic status and homicide
c) Gender and homicide

The word homicide seems to be de rigeur now, rather than just good old, plain Anglo-Saxon murder[211]. I will use it according to my mood. Do not be alarmed.

Before we get all specific about the subject, let's take a look at murder in general.

In 2012, about 437,000 murders[212] were committed across the world. I'll be honest, I thought it would be more. It might well be, depending on any one country's ability or willingness to record such events. It is difficult to judge sometimes what is a killing in war or what is a murder for the sake of it or what is a psychological weapon on the battlefield (AKA the World). The putting down onto paper of such figures is not so strictly adhered to in all parts of the world.

36% of those murders were in the Americas (North and South), 31% in Africa and 28% in Asia[213]. Europe had 5% (so we really are behind in everything) and Oceania just 0.3%. (Oceania comprises: Australia, New Zealand, Fiji, New Caledonia, Papua New Guinea, Solomon Islands, Vanuatu, Guam, Kiribati, Marshall Islands, Nauru, Palau, Cook Islands, Niue, Samoa, Tonga, and Tuvalu).

The global average, that is if you lump all the stats together and pretend the world is just one big country (does this

---

211 ...from Old English - morðor (plural morþras) "secret killing of a person, unlawful killing," **Old Saxon - morth**. www.etymonline.com. I love this stuff!

212 Pedant note: Yes, I know there is a difference - Homicide is the killing of one person by another. Murder is a form of criminal homicide, where the perpetrator intended to kill the other person, sometimes with premeditation (a plan to kill). Manslaughter is another type of criminal homicide. Homicides are criminal, excusable, or justifiable. - *The Legal Difference between Murder & Homicide*. Riccardo Lorenzo Ippolito's Felonies Legal Blogs

213 UNDOC (United Nations Office On Drugs and Crime) Global Study On Homicide 2013

sound like a 1970s Coke advert?), would be 6.2 homicides per 100,000 people. That's the global average. That doesn't mean that each individual country is the same. Indeed, if you look at Southern Africa and Central America, they as regions have over 4 times this rate: 24 murders per 100,000 people. That's utter tomfoolery. South Africa, the Caribbean and Middle Africa have between 16 and 23 murders per 100,000. Eastern Asia, Southern Europe and Western Europe are the regions with the lowest murder levels, you'll be relieved to know and I'm assuming that you're not reading this book in the Sudan. If you are, I have some bad news for you…

What about the way we kill? Well, in the UK we like to use a sharp object more than any other. In the Americas (all of them) guns win hands down. Oceania, which as a region has the fewest murders, likes to use sharp objects, presumably because of the corrupt British genes flowing through them. Asia, however, plumps for 'other', such as poisoning or a damn good beating. Let me chart these for you[214]:

Region	Guns	Sharp Objects	Poisoning, hitting, blunt objects etc
UK	4%	49%	47%
Africa	28%	30%	42%
Americas	66%	17%	17%
Asia	28%	25%	47%
Europe	13%	33%	54%
Oceania	10%	55%	35%
Global Totals	41%	24%	35%

What went against my expectations? Not the Americans, that's for sure. That Asia and Africa had the same amount of gun deaths was surprising. Either Asia has more guns than I thought or Africa has fewer. The impression that I get from the media of Africa is of a continent riven by war and strife, prey to the constant

214 UNDOC Homicide Statistics (2013) and IHME (2012)

crack of gunfire and the ricochet of bullet diplomacy. That Europe should be high on poisonings and beatings? Not really. We still enjoy a good personal service. Globally, it comes as no surprise to find that guns are the big killers. But by only 6%? From 'others'? I was genuinely surprised that sharp objects came below all the alternatives. Live and learn, eh.

Unsurprisingly, men consistently commit more murders than women, in all regions. Only 22% of the murders in the whole wide world (and possibly space too) are committed by women. We shall look at gender differences in murder a bit more later.

Reassuringly, or not, most murders are committed by close family. This means that we don't have to be quite so afraid of strangers as we all seem to be.

Victim/Offender relationship, by victim sex, 1980-2008 (%)		
Victim/offender relationship	**Male**	**Female**
Intimate	7.1	41.5
Spouse	4	21.4
Ex-spouse	0.2	1.9
Boyfriend/girlfriend	3	15.5
Other family	10.9	16.7
Parent	2	4.1
Child	3.6	7.5
Sibling	1.8	1.2
Other family	3.5	3.9
Acquaintance/known	56.4	29.9
Neighbour	1.6	1.7
Employee/employer	0.2	0.2
Friend/acquaintance	46.2	22.8
Other known	8.3	5.2
Stranger	25.5	11.9

It would seem though that it's not just the wife or the hubby of which we should be aware. We are prey to all the family, uncles, cousins, kids – if they share any small percentage of your genes, don't trust them and don't leave them with the knives. And as for friends…Who needs them?

For me, the worse news is not the prospect of my patricide (can I be 'patricided'? Or do I only get to commit patricide?), but the fact that the chance of finding my killer, anywhere in the world, (chances are they'll be in the living room) is really not very good. It's better in Europe, granted, but the rest of the world had better keep their backs to the wall, that's all I can say.

According to NPR, a radio station across the pond, 'If you're murdered in America, there's a 1 in 3 chance that the police won't identify your killer. To use the FBI's terminology, the national "clearance rate" for homicide today is 64.1%. Fifty years ago, it was more than 90%. And that's worse than it sounds, because "clearance" doesn't equal conviction: It's just the term that police use to describe cases that end with an arrest, or in which a culprit is otherwise identified without the possibility of arrest - if the suspect has died, for example.'

Holy cow! I grew up watching Quincy, Columbo, Ironside, Hawaii Five-O (Steve McGarrett has to be the coolest TV cop ever), McMillan and Wife, McCloud and there was not one murder went unsolved. What the hell happened between then and now? Were all these great documentaries untrue?

Even worse than this:

> 'The number of homicides in England and Wales rose by 71 to 574 in the 12 months to September 2015 - an increase of 14% fuelled by rises in knife and gun crime, official statistics show. The police-recorded crime figures include a 9% rise in knife crime and a 4% rise in gun crime, which are thought to reflect a rise in gang violence largely in London and Manchester. The rise in gun crime is the first recorded for eight years and includes a 10% rise in

London. [215]'

The crime-rate is going up. All over the shop, from pick-pocketing to murder. These are based on figures from September 2015. The BBC uses the same figures to bludgeon us into submission.

> 'Murders and killings in England and Wales have increased to their highest level for five years, figures show. The 14% increase in the year to September 2015 was largely due to a high number of deaths in June when 75 people were killed in one month, the Office for National Statistics found. There were 574 murders and killings in total, 71 more than the previous year. Overall reported crime was up by 6% to 4.3 million offences but the ONS said it was due to better recording methods. While the murder rate in England and Wales has risen, it remains significantly lower than it was a decade ago because of 10 years of previous falls... As for murders and killings, the increase cannot be explained by police recording practices, but might be connected to an upturn in the economy, which means more people drinking and getting into fights or a reduction in domestic violence prevention work by cash-strapped police. The increases were mainly in south-east England and Wales. Alternatively, it could be a statistical quirk.'

All those extra dead people might well just be a 'statistical quirk'. That's a comfort. As my head is ruptured by a bullet in the most unpopular mode of murder in the UK, I shall comfort myself with the knowledge that I am but a minority quirk.

In recent years we have constantly been hammered by papers and TV news about the horrendous rise in knife crime.

---

215 The Guardian - *Homicides in England and Wales up 14%*. 21 January 2016.

Neither is there any way to deny the veracity of the figures. Headlines scream that London now more violent than the similarly populated New York. Stabbings in the capital have risen 16 per cent in 2018[216].

> In London, there have been 26 so far this year not including the two deaths on Monday [12 December], compared to a record low of nine last year. There were 16 in 2016, 27 in 2015, 16 in 2014 and 31 in 2013. Children as young as 9 carrying knives amid 'Wild West' violence in UK Among the 119 homicides in 2018, there have been 68 stabbings, 12 shootings and two deaths involving a knife and a gun. A third of the cases involved victims aged 16 to 24. Of these, 30 were stabbed, nine were shot, two died in attacks involving a knife and a gun, and one died in a fall. For the teenagers aged 15 to 19, six were shot and 14 were stabbed.[217]

According to the ONS, not including terrorist events, there was a 14% increase in police recorded homicide offences (from 630 to 719); that is only up to June 2018. Theft, robbery, sexual assault, crimes with violence and public order offences were all on the up. The statistician at the ONS sums it up thus:

> Over recent decades, we've seen continued falls in overall levels of crime but in the last year the trend has been more stable. The latest figures show no change in the total level of crime but variation by crime types. We saw rises in some types of theft and in some lower-volume but higher-harm types of violence, balanced by a fall in the high-volume offence of computer misuse. There was no change

[216] The Sun. Dec 01 2018. Don't worry, it's not just The Sun. I'm not daft!

[217] The Independent. 13 November 2018

> in other high-volume offences such as overall violence, criminal damage and fraud. "To put today's crime survey figures into context, only 2 out of 10 adults experienced crime in the latest year."
>
> Joe Traynor, ONS Centre for Crime and Justice

Is it because people are no longer afraid of the punishments meted out? Or is it, as Simon Webb in *A History of Torture in Britain* says, simply the chance of getting caught which influences the commission of a crime? With a massive reduction of police numbers and a concomitant reduction in police visibility, there has to come a breaking point.

The Guardian on 17 June 2018 wrote:

> Britain's largest police [Metropolitan] force has said it is doing all it can to bring thieves to justice after figures suggested 95% of burglaries and robberies across the UK are not being solved... Scotland Yard said its London sanction detection rates – the way it measures cases that are solved – were 5.5% for burglary and 7% for robbery between April 2017 and April 2018.

In the Telegraph on 13 April 2018, Metropolitan Police commissioner Cressida Dick

> 'when asked whether she was concerned that the murder detection rate fell to 72 per cent in London last year from its normal levels of around 90 per cent she said that officers faced a "challenging environment...Not only in terms of the volume," she said. "In fact, a bigger issue is the complexity. These cases are not a classic whodunnit where somebody is deceased and we have no idea who has done that. More often than not we have a very good idea very early about who was involved in that fight

> on the street. Proving which one of those people did that is hard, proving that when we are met very frequently by a complete wall of silence, where very often nobody wants to tell us anything initially. Proving to [a] standard where CPS will charge is a very big challenge.' She argued that a detection rate above 70pc was still "strong and good and great"...Commander Jim Stokley, head of the Met's Gangs and Organised Crime Command, added that the "age demographic is changing", pointing out: "The victims are getting younger and younger and the suspects are getting younger.".

Of course, that statement of a 'challenging environment' is open to interpretation, but should there exist a situation where police faced a 'challenging environment' in the first place and, if that challenge exists, should the resources to take on that environment also exist?

a) Race and Homicide

> 'What we are dealing with is not a general social disorder; but specific groups or people who for one reason or another, are deciding not to abide by the same code of conduct as the rest of us… The black community – the vast majority of whom in these communities are decent, law-abiding people horrified at what is happening – need to be mobilised in denunciation of this gang culture that is killing innocent young black kids. But we won't stop this by pretending it isn't young black kids doing it.'
>
> Tony Blair, 2007
>
> 'The Commission for Racial Equality broadly backed Mr Blair, saying people "shouldn't be afraid

to talk about this issue for fear of sounding prejudiced".'

The Guardian. 12 April 2007. *Blair Blames Spate of Murders on Black Culture.* Patrick Wintour and Vikram Dodd.

In 2007, in London, the greatest attribution of knife crimes between the ages of 18 and 25 years were to:

Whites - 61% Total – 3,372
Blacks - 20% Total - 1,106
Overall Total - 5,528

That is a ratio of 3.04:1 when comparing white knife crime to black knife crime in that age bracket. The population of London in 2007 was 7,556,900. 10.6% of these were black.[218]. That is 755,690.

In 2011, the year of the last census, 4,887,435 or 59.79% were white. The total of Black/African/Caribbean/Black British was 1,088,640 or 13.32%[219], up by 2.7% on 2007. There were 4.4 times as many whites as blacks in London in 2011. For those of you who, like me, hate wading through figures, here is the above in a table:

Ethnicity	Number	% of Total
UK Population[220]	63.2 m	
England Population	53 m	84.12%

[218] Young Black People and the Criminal Justice System
[219] 2011 Census: Ethnic group, local authorities in England and Wales". Office for National Statistics. 11 December 2012.
[220] Census information scheme GLA Intelligence. Diversity in London, June 2013.

London Population	8,173,941	15.42%
White	4,887,435	59.79%
Black/African/Caribbean/Black British	1,088,640	13.32%
Asian	1,511,546	18.49%
Mixed Race	405,279	4.96%
Other	281,041	3.44%

Now in 2010, the year before (the closest available figures to the last census), 53% of knife crime in the specified age bracket was committed by whites and 23% by blacks.

Proportionally, there is little to choose between these figures and yet the news was full of the murderous black tide sweeping across our nation and this news, forgive me, coloured our perceptions.

Number of Knife Crimes That Occurred in London Between 2005 and 2007 Involving Persons Aged 18 - 25 and The Percentage Breakdown. . D. Lindo 13 July 2011.						
YEAR	**2005**	**2006**	**2007**	**2008**	**2009**	**2010**
ASIAN	15%	18%	17%	19%	21%	23%
BLACK	18%	21%	20%	20%	21%	22%
EAST ASIAN/SE ASIAN	1%	1%	1%	1%	1%	1%
MIDDLE EASTERN	0%	1%	0%	0%	1%	1%
WHITE	65%	59%	61%	59%	56%	53%
UNKNOWN	0%	1%	0%	0%	1%	0%
TOTALS	4673	3956	5528	4737	1987	1575

The number of knife crimes in 2007/8 in South Yorkshire was 381. In other areas, the knife problems were even worse:

AREAS WITH HIGHEST KNIFE CRIME INCREASES			
	2002	**2004**	**RISE %**
Nottinghamshire	338	650	92
Bedfordshire	79	113	43
Devon & Cornwall	108	152	41
Lincolnshire	402	497	24

Why was this not up in lights across the nation's skies like a nuclear *Aurora Borealis*? Because much of it was committed (whisper, whisper) by white people.

So how mad was Tony Blair being? Not too mad really. His words can be interpreted as racist by some, but the CRE gets him out of that scrape. What he was doing was reacting; politicians do that. He was reacting to the fact that, 'the capital saw a significant rise in 2007 when 26 youngsters were killed, up from a stable average of 17 a year since 2000.[221]'

The problem was that it was happening in London and the media and politics is notoriously Londoncentric. None of the above figures in lesser places made headlines. Nottinghamshire's violence is legendary up here. The papers don't shout about that and Tony Blair certainly didn't seem to give a damn about it.

The Prime Minister wasn't being prejudiced, he was being *isolationist*, in financial terms, in terms of social status, in terms of postcode and probably politically too. Maybe in terms of colour also.

The other, possibly the main thing, that made people stand up and pay attention, was the colour issue. This country is revoltingly xenophobic; it always has been, right back to medieval times. We always needed someone to blame and if it wasn't the Jews, it was the Flemish and if you couldn't find either of those, then it was always the Frenchies' fault (the Spanish came next,

[221] http://news.bbc.co.uk/1/hi/uk/7777963.stm

followed by the Dastardly Dutch).

The question is not so much as…no…wait a moment…there was no question. It was just a barebones statement repeated over and over in order to sell papers and gain votes. The question *should have been*: Why? Why was there a concentrated area of Britain's capital city that was suddenly grabbing the news in the worst possible way?

We'll look at that later, but I suspect it had something to do with opportunity, money, education, bias, prejudice and culture.

HOMICIDE TYPE BY RACE 1980-2008 – VICTIMS (%)			
	WHITE	**BLACK**	**OTHER**
INTIMATE	55	42.7	2.4
FAMILY	59.2	38.2	2
INFANTS	56.2	28.6	2.8
ELDERS	69.6	28.6	1.8

HOMICIDE TYPE BY RACE 1980-2008 – OFFENDERS (%)			
	WHITE	**BLACK**	**OTHER**
INTIMATE	54.2	43.5	2.3
FAMILY	59.2	38.3	2.5
INFANTS	55.8	41.6	2.6
ELDERS	56.3	41.9	1.8

HOMICIDE TYPE BY RACE 1980-2008 – OFFENDERS (%)			
	WHITE	**BLACK**	**OTHER**
Felony murder	38.4	59.9	1.7
Sex-related	54.4	43.4	2.2
Drug-related	33.2	65.6	1.2
Gang related	53.3	42.2	4.6
Argument	47.5	50.2	2.4

Workplace	70.8	25.8	3.3

The tables, above[222], is from the Good Ole (US of A) and breaks down murder by race between 1980 and 2008. It is worth remembering how much some parts of the US have changed. New York in 1980 was just about to come out of the tailspin in which it had found itself. It had become a no-go city, its worldwide reputation apparently tarnished beyond repair by the high crime and drug levels. It is a wonder that people who could afford it didn't just evacuate and leave it to those who couldn't afford to leave, leaving behind a John Carpenter *Escape From New York* scenario. If you haven't seen that film, a) that will mean nothing to you and b) what is the matter with you? Watch the damned film. Kurt Russell growls like a Rottweiler with piles and Harry Dean Stanton's in the film. What more could you want?

I digress.

The age-old story of you're more likely to be killed by a family member still holds strong. It always has. In both the victim and the offender categories, the whites have higher figures. Why? This is half the problem with these charts, you can say anything by interpreting them however you want because they are, due to the enormous nature of the figures, vague. My guess would be that there are fewer black 'in-house' murders because as family units, they are tighter. I would also say that poverty brings them together. It can be a very strong glue. Look at the drug-realted events, though. Blacks are almost twice as high. They are trapped by culture, poverty, lifestyle and peer pressure. You would think that gang-related killing would be about the same, but they are not; white gang-related murders are a quarter as much again. What are the whites trapped by? Money and boredom? Why? Well, 84% of white victims were killed by whites and 93% of black victims were killed by blacks. The first thing that comes to mind is that the white gangs are either a) more numerous and b) more 'successful'.

I am assuming here, of course, that there are no white

[222] *Homicide Trends in the United States, 1980-2008 Annual Rates for 2009 and 2010.* U.S. Department of Justice. November 2011.

people out of work, none that are subject to abuse by family and none facing peer pressure every day, the same things that their black counterparts face. That would be a ridiculous thing to say, but

> 'blacks were disproportionately likely to commit homicide and to be the victims. In 2008 the offending rate for blacks was seven times higher than for whites and the victimisation rate was six times higher… It's true that around 13 per cent of Americans are black, according to the latest estimates from the US Census Bureau. And yes, according to the Bureau of Justice Statistics, black offenders committed 52 per cent of homicides recorded in the data between 1980 and 2008. Only 45 per cent of the offenders were white. [223]'

Let us not forget however, that this is taken over a twenty-eight year period and presents averages, not year on year. Neither is it just about gangs. If we take a look at gangs, the culture and how it affects the individual by race, the picture tends to lean a little to the left. There is one area that gets scant attention in the table and that is 'other'. According to the others 'column', those 'others' contribute very little to the national naughtiness.

You would be forgiven for thinking that whites and blacks were the only ones who can be bothered to get off their backsides and commit murder.

There is also the false truth of our assumptions, created by our own prejudices, our own biases, our own cultures and our own socioeconomic standing. We expect the blacks to be the main cause of crime. That is what the media tells us. If you look at the news from America right now, we are seeing disturbances made by black communities in response to the seemingly carefree shooting of blacks by police. There is a slant in the media which

---

[223] FactCheck: do black Americans commit more crime? November 27, 2014. By Patrick Worrall

makes the viewer feel that the blacks should actually behave themselves and not mucky up the streets protesting. But there is also a part of us, certainly in me, which knows that they are right, that their anger and frustration is justified after years of financial, racial and legal oppression

However, not all is quite as it seems. Of course those figures on the '1980-2008' chart are right, but take a look at the chart below from 2011[224], which focuses on gangs. The way the figures are broken down suggests a whole different picture, a subculture divided by turf, race, religion, family and tradition.

Average Race and Ethnicity of Gang Members (%)				
	Larger Cities	**Suburban Counties**	**Smaller Cities**	**Rural Counties**
Black/African American	39	32.7	20.3	56.8
Hispanic or Latino	45.5	51	53.8	24.8
White	9.7	9.1	14.6	14.9
Other	5.8	7.2	11.3	3.4

Look who takes top billing now (except rurally). Not blacks, not whites, but Hispanic and Latino. So why the discrepancy between this chart and the '1980-2008' chart? Because there are an estimated 33,000 gangs in the USA. Many of them are white. More than I ever thought before doing the research for this book (I too came at it with my own bias and media influence). It isn't until you break these down into the small communities that you begin to get an idea of the differences.

I used to work for a really racist, homophobic, misogynistic prat who said that the blacks and foreigners were the cause of all the violence/crime in Britain. My argument to that was, not in Barnsley. Most of the crime in Barnsley is by people of his pasty-faced status (and with his moral standing too). We actually have a very low black/Asian population in Barnsley and

224 www.nationalgangcenter.gov/Survey-Analysis/Demographics

even if there was, it doesn't necessarily follow that they would naturally be the epicentre of any trouble. It was only by breaking down the demographic that you got a true figure. Sure, if you went to a predominantly black area of London, you would find that a lot of the crime was by black people. You might also find a lot of the cultural deficits that you would find in those gangs in the Good Ole. This isn't giving excuses. I despise crime and criminals, anyone who picks on weaker people, but they are *reasons.*

Year	**Hispanic Or Latino**	**Black Or African American**	**White**	**All Other**
1996	45.2	35.6	11.6	7.5

Then if we go back to 1996, they are still top of the tree.

> 'In some neighborhoods, many members of a family have belonged to the same gang. These multigenerational gangs develop in different settings, but have been most often observed among Hispanics. Sanchez-Jankowski (1991) reported that many gang members told him that their families had a long history of gang involvement that included older brothers, and in a considerable number of cases, fathers and grandfathers. Thirty-two percent of the Los Angeles fathers he interviewed said that they had been members of the same gang to which their children now belonged, while 11 % reported that four generations of their family had membership in the same gang… According to Sanchez-Jankowski (1991), tradition plays an important role in multigenerational gangs. He argues that the long history of multigenerational gangs, coupled with parents' former involvement with the same neighborhood gangs, brings a sense of tradition to the gangs… As indicated in a comment by a gang youth Sanchez-Jankowski

interviewed, many youths in these neighborhoods feel that their families and community expect them to join a gang:

> I joined because the gang has been here for a long time and even though the name is different a lot of the fellas from the community have been involved in it over the years, including my dad. The gang has helped the community by protecting it against outsiders so people here have kind of depended on it…This will help me to get help in the community if I need it some time.[225]'

This suggests a culture, based upon race and reinforced by peer pressure, family pressure and lack of opportunity. It doesn't suggest vast wealth gained by nefarious means, just a way to find those basics that we all need to survive. The article, which is grim yet fascinating reading, goes on to say:

> 'Many gang members report that they live with parents or stepparents who are alcoholics, chronic drug-users, abusive (physical, sexual, and emotional), or involved in illegal activities. These conditions create considerable stress that youths may try to alleviate by joining a gang. Over a quarter of the women in Moore's (1991) study of East Los Angeles gangs reported that a family member had made sexual advances while they growing up. Almost a quarter of the men and about half of the women resided with a heroin addict during their childhood, and about half of the men and over half of the women had a member of the household die during their formative years.'

---

[225] www.encyclopedia.com/topic/Gangs.aspx. *Gangs*. *International Encyclopedia of Marriage and Family*. 2003 COPYRIGHT 2003. The Gale Group Inc.

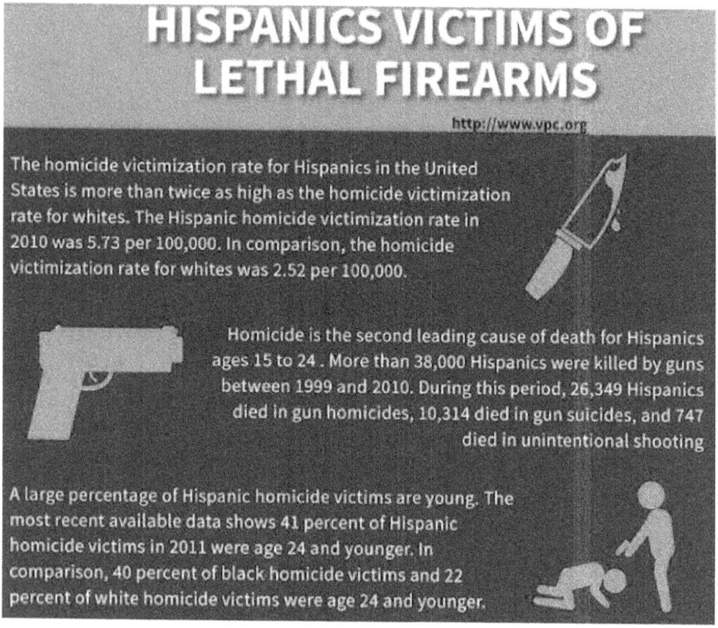

HISPANICS VICTIMS OF LETHAL FIREARMS
http://www.vpc.org
The homicide victimization rate for Hispanics in the United States is more than twice as high as the homicide victimization rate for whites. The Hispanic homicide victimization rate in 2010 was 5.73 per 100,000. In comparison, the homicide victimization rate for whites was 2.52 per 100,000.
Homicide is the second leading cause of death for Hispanics ages 15 to 24 . More than 38,000 Hispanics were killed by guns between 1999 and 2010. During this period, 26,349 Hispanics died in gun homicides, 10,314 died in gun suicides, and 747 died in unintentional shooting
A large percentage of Hispanic homicide victims are young. The most recent available data shows 41 percent of Hispanic homicide victims in 2011 were age 24 and younger. In comparison, 40 percent of black homicide victims and 22 percent of white homicide victims were age 24 and younger.

The reasons that people join the gangs are multifactorial, but there is always a commonality, regardless of race, that drives people towards the gang culture:

- Profiting from organized crime, which could be a means to obtain food and shelter, or access to luxury goods and services
- Protection from rival gangs or violent crime in general, especially when the police are distrusted or ineffective
- Personal status and coolness
- A sense of family, identity, or belonging
- Intimidation by gang members or pressure from friends
- Family tradition

- Excitement of risk-taking
- Lack of parental supervision
- Family instability
- Family members with violent attitudes
- Being part of a socially marginalized group (e.g. ethnic minority)
- Family poverty
- Lack of youth jobs
- Academic problems (frustration at low performance, low expectations, poor personal relationships with teachers, learning disability)
- Violent crime committed by others against the potential gang member, or friends or family
- Involvement in non-gang illegal activity, especially violent crime or drug use
- Low self-esteem
- Lack of role models
- Hyperactivity
- Early sexual activity
- Illegal gun ownership[226]

These apply equally to the UK as they do to America, much of it is down to human nature and the need to feel safe, a sense of belonging, a sense of family when it cannot be found elsewhere. But we all did that as teenagers, just not to that degree. The greater the lack of identity, the greater the search for it. If we look at

226 *"Why Young People Join Gangs"*. lapdonline.org. *"Gang Alternatives Program (GAP)"*. gangfree.org. *"Why Do Youth Join Gangs?"* ojjdp.gov. "*Gangs give members sense of purpose, belonging – but there's a price"*. northwestern.edu. "*Why Young People Join Gangs and What You Can Do"*. violencepreventioninstitute.com. *"Gangs give members sense of purpose, belonging – but there's a price"*. News.medill.northwestern.edu. 2010-05-20.

Maslow's Hierarchy of Needs, these motives for joining a gang

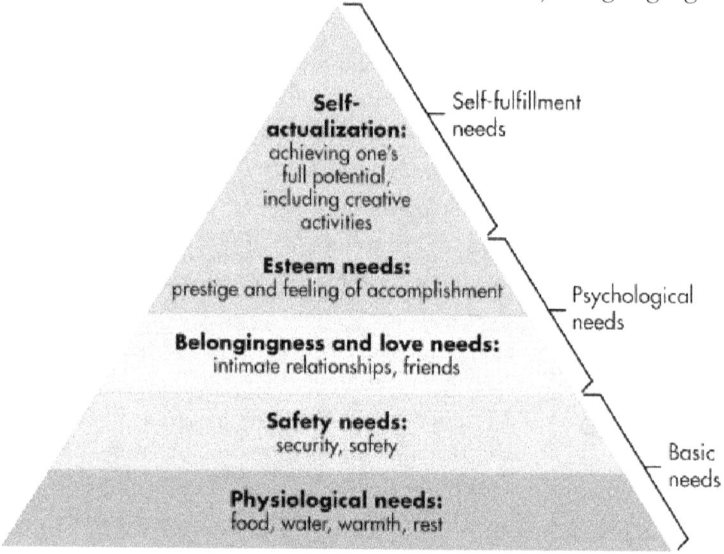

pretty much fit neatly into those categories on the hierarchy.

Self-actualisation for those who live within the gang culture might not actually be the same definition as yours or mine. We all come at this life from very different angles.

So where does this leave us in the debate of race and homicide? Confused? Hell, yes. More aware? Yes. We could look at another set of statistics, such as murders in the workplace, and come out with a whole new series of conundrums and hypotheses. The important thing is context.

US DEMOGRAPHICS BREAK DOWN 2010		
Race/Ethnicity	**Number**	**% US Population**
None-Hispanic White	196,87,552	63.7
None-Hispanic Black	37,685,848	12.2
Non-Hispanic Asian	14,465,124	4.7

Non-Hispanic Native American Or Alaskan Native	2,247,098	0.7
Non-Hispanic Native Hawaiian Or Other Pacific Islander	481,576	0.2
Non-Hispanic Some Other Race	604,265	0.2
Non-Hispanic Two Or More Races	5.966,48	1.9
Hispanic Or Latino	50,477,594	6.3
Total	308,745,538	100

The US population in 2010 was 308,745,538. The demographics break down as shown in the chart left.

So, Whites make up 63.7% of the population, blacks, 12.2% and Hispanic/Latino, 16.3%.

According to the American Department of Justice, blacks made up 52.4% of homicide offenders while whites accounted for 45.4%. 'Other,' or Hispanic/Latino, criminals came in at 2.3%. The offending rates for blacks was almost 9 points higher than whites.

Put these two pieces of information together and you end up with this:

Ethnicity	% of Population	% of homicides
Blacks	12.2	52.4
Whites	63.7	45.4
Hispanics	16.3	2.3

Yep, we are back where we started because we have focused on *all* homicides and not just a particular section of homicides within a particular section of society (these figures were taken from a different source[227] this time). However, those

[227] https://en.m.wikipedia.org/wiki/Demography_of_the_United_States. www.census.gov.

sections of society which are most deprived or excluded - socially, financially, academically, politically – that live in isolation from the rest of society, in economic and social enclaves, might be the ones that push those figures up. It is possibly/probably more indicative of the society in which they live rather than their colour/ethnicity. Hispanics, blacks and other races are not, so far as we know, genetically inclined to crime any more than white people.

And yet…whites commit 70.8% of murders in the workplace compared to 25.8% of the blacks. Is it because the black ladies and gentlemen are so much easier to work with or because white collar jobs, which attract more white people, have a higher cull rate as a refreshing part of staff turnover?

In the UK, the gang culture is slightly different, although it clearly exists. According to the Home Office paper, *2010 To 2015 Government Policy: Knife, Gun and Gang Crime*, 'Gang members carry out half of all shootings in the capital and 22% of all serious violence.'

The problem that has arisen in recent years is the open accusation that the British police is institutionally racist and that, because of its perception of blacks and Asians, targets them more frequently than they do white people. If you target one section of society more often than another, it stands to reason that you will find more wrong[228].

> 'The biggest disparity emerged in Dorset, where 200 stop and searches were carried out on black people between December 2014 and April 2015. According to the latest census figures, the county is home to just 3,200 black people.
> Over the same period 2,549 stop and searches were conducted on Dorset's white population of 714,600, making it 17.5 times more likely that a black person was targeted. The force partly attributed the figure to operations against drug gangs and to its black population being

[228] Image: *Ethnic Groups England and Wales, 2011* - www.ons.gov.uk

> underestimated... The ratio was 10.5 in Sussex, while other high figures were recorded in Norfolk (8.4), Warwickshire (7.6) and Surrey (7.3). In London, a black person was 3.2 times more likely to be stopped and searched than a white person, with slightly lower ratios in the large urban forces of the West Midlands (3.0), Greater Manchester (2.6) and South Yorkshire (2.1).[229]'

17.5 times more likely to find something wrong in Dorset apparently.

But do the figures fit the suspicion?

In certain areas yes - and, just to add a splinter to my backside from this rotting fence upon which I sit, in other areas, no.

> 'The White ethnic group accounted for around three quarters of the total number of homicide victims in the last three years – although the total number of homicides has decreased over this period. However, the rate of homicide…was 4 times higher for Black victims compared with White victims, and 1.5 times higher for Asian victims. These rates were broadly similar when comparing London (which is more ethnically diverse) to the rest of England & Wales[230]…"Just because the police treat black men as more criminal than white men, it does not mean that they are." Simon Woolley, speaking as the director of the Operation Black Vote pressure group, but who is also a commissioner on the Equality and Human Rights Commission, said: "Although the charge rates for some criminal acts amongst black men are

---

[229] *Black People Still Far More Likely To Be Stopped And Searched By Police Than Other Ethnic Groups*. The Independent. 6 August 2015.

[230] *Statistics on Race and the Criminal Justice System 2014*. Home office.

high, black people are more than twice as likely to have their cases dismissed, suggesting unfairness in the system.[231]'"

The suggestion of unfairness in the system is a very heavy accusation to make. It implies that the law is not impartial. The idea that the law is neither colour-blind nor sensitive to the shortcomings and indignities of poverty, is not biased towards which school the defendant went to or in which area the defendant lives or how much he earns, causes the worm inside me to shuffle uncomfortably. It's not a pleasant sensation.

Yet if we look at those who are involved in the law, the practitioners of our justice system, we do find some interesting figures:

Only 12.7% of the total practitioners in our justice system come from black, Asian or ethnic minorities. It would be interesting to know the socioeconomic status of the remaining 87.3% white remainder. I am not saying that there is a deliberate bias in our legal system, but I do question the susceptibility of the system to unintentional bias based upon perceptions and financial and social segregation from those that they sentence. 94% of the judiciary are white. 92% of magistrates are white. 94% of police officers are white. 81% of the Criminal Prosecution Service are white.[232] These are 2014 figures. Four years later '12 per cent of magistrates are from black, Asian or minority ethnic backgrounds'.[233] 'As at 31 March 2018, 93.4% of police officers were white, and 6.6% came from all other ethnic groups'.[234] Black and ethnic minority members of the judiciary had increased by 1% to 7% by 2017.[235]

Not so much has changed then.

---

[231] *Violent Inner-City Crime, The Figures, And A Question Of Race.* Daily telegraph, Sunday 02 October 2016.

[232] *Statistics on Race and the Criminal Justice System 2014 England and Wales.* Ministry of Justice.

[233] Jane Croft, Financial Times. August 21, 2018

[234] Gov.uk: Police workforce

[235] Courts and Tribunals Judiciary

Those perceptions are important, because they form the basis of our opinions and can create a mythical reality that is a long distance from actual reality.

The figures that I *really* wanted, how many blacks, whites, Asians and 'others' were convicted of murder (murder by race and ethnicity in other words) were very difficult to come by. We seem more reluctant to share than the Good Ole.

We know that the rate of homicide was 4 times higher for black victims compared with white victims. We know that the proportion of blacks in prison over the age of fifteen is higher than whites and way higher than other ethnicities. We know that reoffending is higher in blacks than whites, though not by much. We know that police activity is geared more towards blacks than whites and that the employees in the legal system are predominantly white, but I'm buggered if can find anything about who commits the most murders. Bet I can have a good guess.

Sorry.

I hope you found it interesting anyway.

The conclusion to the article from www.encyclopedia.com, which I referred to earlier, goes someway to presenting a final view on this particular problem:

> 'Gang life and family are fraught with contradictions. The family is idealized as a place for nurturance, support, and protection. However, the majority of gang youths come from families under severe strain; families often unable to provide these things. The gang is often demonized (not without reason) as a source of delinquency and violence. Yet gangs also act as an important source of support for these youths, compensating for what is lacking in their home life. Gangs can act as surrogate families to youths, providing a sense of belonging, identity, status, and protection. Youths who do not receive these things from family or other social institutions may seek them elsewhere, and in the socially disorganized neighborhoods

where gangs exist, they are an alternative option.'

If you expand this, you could apply it to almost any aspect of life and to any race, anywhere, and not just gangs. Our mentality at work, think of those hideous cliques, though veneered by social refinements, is really no different, regardless of status and colour. It's the same with school, the football terraces or the pub – human nature will out and it will usually out in the most unsettling ways. Just don't be fooled by a tie and a nice pair of shoes.

This all leads us nicely into...

b) Socioeconomic status and homicide

> 'The most striking and consistent relationship between poverty and crime is that found between inequality, poverty and homicide. This relationship holds across many different settings - among developed and developing countries, both between and within countries[236].

Our socioeconomic status is at the back of our minds all the time, creeping forward like a disabled, cobwebbed ghost in the attic, struggling for attention. We constantly reflect on how something that happens reflects upon us as an individual. One obvious example is that the tutors I have worked with are results driven. If, at the end of the year, they get a high failure rate, they lose face and possibly position. The stress that this produces is phenomenal and, in my honest opinion, quite unreasonable. In fact, the stress might actually be stopping them carrying out their jobs properly, leading to those poor results.

What about something more immediate? Somebody in class is constantly pissing about – on their mobile, chatting to their mates, gazing out the window, generally being a distraction. What does the teacher do? Does he attempt to maintain discipline and

---

[236] Joseph Rowntree Foundation. *Anti-Poverty Strategies for the UK. Poverty and Crime Review*. May 2014.

risk losing face when told to piss off? Or is he prepared to take it to the limits, whatever the cost, in order to maintain discipline and give those in class who want to learn the chance to do so?

His status in that small society is immediately up for grabs. What he does next might well follow him for the rest of his career. If it all blows up in his face, the students will never respect him again, it would get back to the management, be a black mark on his record and ruin his chance of becoming Head of Department. He might lose his job if he handles it so badly that it leads to a complaint. Loss of job equals loss of income. You're only two wage packets away from being on the streets.

It's actually much easier to do nothing.

The May 2014 report, *Anti-Poverty Strategies for the UK. Poverty and Crime Review,* by the Joseph Roundtree Foundation is a wonderfully common-sense snippet of information examining the links between poverty and crime[237]. It doesn't leap to conclusions or scream hysterically from the rooftops. It just says. It's nice when things just say.

So what did the report share with us? It found that:

- The most striking and consistent relationship between poverty and crime is that found between inequality, poverty and homicide
- The consequences of poverty such as homelessness and poor housing and disrupted families can lead to frustration and violent aggression.
- The cycle of humiliation – the financial humiliation among violent offenders, leads them to humiliate others, either because they feel they have earned the right to do so or because it makes them feel better about themselves – it's an ego boost.

---

[237] Colin Webster and Sarah Kingston. Centre for Applied Social Research. Leeds Metropolitan University

- 'Violence as a cultural reaction to economic hardship' can become the norm when dealing with the problems of everyday life. It gains respect and status and produces fear in others, removes personal humiliation from the offender and might then be passed on to children as a mode of normal behaviour.
- Small reductions in income inequality cause large reductions in homicide.
- Inequality influences homicide, whereas a society's average income level does not.
- Levels of absolute poverty were significantly associated with higher rates of serious violent crime.

The report, in fairness, also stated that the relationship between income levels and homicide rates was less significant than ethnicity, social class or non-economic factors. However,

> '…violence is one of the few options available to those without the economic means to deal with problems or cites of everyday life. Absolute deprivation may also produce emotional situations which escalate into violence…simply put, the absolute deprivation approach suggests that violence can occur among such individuals because everyday life is difficult.' [238]

So it is the disempowering of the individual due to financial circumstances, the social inequalities that come with this and the feelings of humiliation and powerlessness that force the individual to take up arms in an attempt to regain some of that power. This is skewed thinking, but it belongs to a skewed situation and if the skewed situation becomes the norm, then so

---

[238] Parker RN. 1989. *Poverty, Subculture of Violence, and Type of Homicide.*

do its consequences. People need a sense of worth, without it, they turn against the society that they perceive has turned against them. In other words, money does matter, but the combination of ethnicity, social class and economic factors increases the fallout.

Economist John Harsanyi gives this idea credibility and depth when he says that "'apart from economic payoffs, social status seems to be the most important incentive and motivating force of social behavior." The more noticeable status disparities are, the more concerned with status people become.[239]'

> 'Psychologist PJ Henry at DePaul University recently published an article demonstrating that low-status individuals have higher tendencies toward violent behaviour…individuals from low-status groups (e.g. ethnic minorities) tend to engage in more vigilant psychological self-protection than those from high-status groups. Low-status people are much more sensitive to being socially rejected and are more inclined to monitor their environment for threats. Because of this vigilance toward protecting their sense of self-worth, low-status individuals are quicker to respond violently to personal threats and insults.[235]'

So who you are, where you come from, the colour of your skin and what you earn all play a significant role in the individual's leaning towards violence.

The consequences of those acts are also affected by your social status and income.

Peggy O'Hare in *Study Figures Odds of a Killer Getting Death,* in the Houston Chronicle on March 4 2010, was in no doubt that crime, punishment, race and ethnicity were linked, all the way to the gas chamber. What she found was:

---

[239] www.scientificamerican.com/article/the-psychology-of-social/. Adam Waytz. December 8, 2009. *The Psychology Of Social Status: How The Pursuit Of Status Can Lead To Aggressive And Self-Defeating Behaviour.*

- The death penalty is more likely to be imposed on convicted murderers who kill whites or Hispanics who have college degrees.
- If a victim had a college degree, the odds of prosecutors seeking death were 1.5 times higher and the death sentence was more than twice as likely to be imposed.
- Convicted capital murderers also were more likely to get the death penalty when defended by court-appointed lawyers.
- Those who hired attorneys to represent them for the entire case were never sentenced to death.
- Convicted capital murderers were six times more likely to get a death sentence when they killed married whites or Hispanics with college degrees and no criminal record - as opposed to unmarried black or Asian victims with records and no college degrees.
- Prosecutors seeking a death sentence were 1.8 times higher if the victim was white or Hispanic
- The death sentence was twice as likely to be carried out if the victim was white or Hispanic.
- Prosecutors seeking a death sentence were 1.8 times higher if the victim was white or Hispanic.
- If the victim killed had no criminal record, the district attorney was 1.8 times more likely to seek a death sentence against the accused.[240]

So, if you could afford to hire an attorney, you stood more chance of getting off. If you were a white or Hispanic victim, your assailant would be more likely to be more harshly punished. If your victim had a good education and no criminal background, you

---

[240] According to a new study that examines 504 Harris County capital murder cases that occurred between 1992 and 1999.

were really down the pan.

Wealth, education and status all impacted upon the outcome of a court case and, ultimately, whether the offender lived or died.

This is a stunning indictment of the US (and I don't doubt the British) criminal system. Impartiality goes out of the window when the nuances of our personal psychology come into play. It might not be intentional, but it's there, perhaps because of your upbringing, the area in which you work, the neighbourhood in which you live and your race and ethnicity.

Obviously, this is not as in-depth as I would ideally like to go. Harris County, where the above study was made, does not represent the world. It is in the southern states of the USA.

> 'Crime has long been associated with poverty, and many of America's states with the lowest per capita income are in the South - including Mississippi, West Virginia, Arkansas, Alabama, and Kentucky.[241]'

The southern states of the USA have the highest murder rates in America. People might be reacting the best way they know how and in the way they have done for generations. It is interesting to note though that among the ten states with the lowest murder rates, six of them no longer have the death penalty.

> 'On November 10 the Justice Department released its annual Uniform Crime Report for 2013. The report revealed an overall decline of 5.2% in the national murder rate. The Northeast had the lowest murder rate--3.5 murders per 100,000 people - and the sharpest decline from last year. The South again had the highest murder rate (5.3). The West had the second-lowest murder rate (4.0), followed by the

---

241 *Why The South Is More Violent Than The Rest Of America.* Erin Fuchs. Sep. 18, www.businessinsider.com.

> Midwest (4.5). The states with the highest murder rates in the country were Louisiana (10.8) and Alabama (7.2). The states with the lowest rates were Iowa (1.4) and Hawaii (1.5). The Northeast has also had the fewest executions in the modern era, with 4, and none since 2005. The South has had the highest number of executions (1,132) since 1976. The average murder rate for states with the death penalty (4.4) was higher than the average rate for states without the death penalty (3.4).'

Could it be that Harris County is clinging on to age-old tried and trusted methods of dealing with such grave problems without turning an eye towards the rest of the world? Is the rest of the world, in its attitudes to equality and diversity, benefitting from its greater appreciation of the nature of humanity? There is no doubt that tradition plays a massive part. We have seen that gang membership can pass from generation to generation, so why not prejudice and bias? It isn't until someone comes in with a statistical crowbar and a good deal of stamina and guts that approaches change. Look at the race problems of the 60s in the US. I sure as hell wouldn't have had the guts that some people had to stand up to the bigots and that included the institutionally racist authorities.

And don't let's get all high and mighty here in the UK. We are still human and still very prone to the same human foibles that affect our cousins across the world. I'm not going to tar and feather the Americans with this. They are actually very forthcoming and generous with their statistics and have certainly made this job easier than the UK authorities with ease of access to information.

The Americans have to a great degree faced up to their (many and ingrained) problems and are at least trying to right them, even as I write. I'm not sure that we are doing the same – the extreme measures that we are going to at the moment in order to suppress opinion in the name of political correctness by cultural bullying is actually as bad as the racism and homophobia that we displayed in the 60s and 70s.

c) Gender and homicide

More men are murdered than women. That is a stone-cold fact. I thank God that I'm not aged between 15 and 29 anymore - I am past my most vulnerable age. The testosterone that used to drive me and caused me to constantly bang my head against the wall in the sincere belief that it was doing more harm to others than to myself, is draining away like dirty oil from an old engine sump. My wife, I am happy to say, is more dangerous than I am. I don't know that through any statistics. I just know that.

Men actually lead the field as both victims and murderers. I don't say that with any pride. I would rather be as good at multi-tasking as women and have the ability to cook without the food being blackened or coming out colder than when it went in. In deference to my wife, she is better at practically everything than me. She's probably a better murderer too. Who knows?

In the year ending March 2015, in the UK two thirds of the murder victims were male. Where it gets considerably more dangerous for the ladies is, sad to say, in the home. 77% of female domestic murder victims were killed by their husband/partner. The remaining 23% were killed by a family member. I think that's quite frightening. Out of 551[242] murders in England and Wales in 2012/13, 77 (14%) of these were by their partner/ex-partner, as compared to 16 (3%) males. Half the female homicides in 2012 (47%) were killed by their partners or family members. 97% of female domestic homicide victims were killed by men. The highest of these numbers was to be found in Asia, then Africa.

---

242 Home Office / Office for National Statistics

Globally, 79% of murder victims were female[243].

The recession hasn't helped. At the risk of sounding like I am making excuses, which I most assuredly am not, the massive loss of jobs and income caused by the affair, with the concomitant dependence by males upon females as a source of income and strength, has led to the emasculation of men, to a change in their role, to them playing second fiddle both financially and psychologically to their wives/partners.

This builds up resentment and feelings of inadequacy in the men, while at the same time perhaps giving the women (a feeling of) power and freedom that they previously had not had and leading to the first available expression of frustration by men, violence; the stresses and challenges brought to masculine identity by changes enhancing women's position of women has led to increased levels of violence against them[244].

With regard to race, black women were 3 times more likely to be murdered than a white woman, whereas for the men it was eight times greater in blacks than whites. *Eight times*. Not only does being male increase your chance of being extinguished, but you then have to worry about colour and, with the European mess we are in lately, there is now an increase in crimes against white male immigrants such as Poles.

When it comes to offending, men offend more than women. 95% of murders in the world are by men.

Motives are different though. Women kill for more prosaic reasons.

---

243 Homicide and Gender 2015 – UNDOC

244 Joseph Rowntree Foundation. *Anti-Poverty Strategies for the UK. Poverty and Crime Review*. May 2014.

> '...women made rational choices based on the options they had available and their economic situation. The attractions of crime for these women were an alternative to the indignities, humiliations, delays and frustrations of claiming welfare.[241]'

It is also offered that women kill for a much more basic reason – they like shopping, shoes and shiny things. The Joseph Roundtree report goes on to say:

> 'Others have argued that female crime rates are less driven by poverty than a desire to engage in consumer society. The inability of poorer groups to legitimately attain consumer items considered desirable, and therefore use illegal methods of attaining them, is nevertheless still about being poor in a society where the attainment of certain good is the norm. (Box and Hale, 1984)'

Any woman driven to murder for a SMEG fridge freezer has to have a certain twist in their thought processes, let's be honest, but the pressure today, upon all members of society, particularly the female and the young, drowning in the constant high tide of advertising and the subsequent *inter alia* competition, is ridiculous.

So, what about the methods? Do men and women differ in their preference for a good topping? Yes, they do, but the whole women and poison thing is a bit of a myth, unless it is self-poisoning.

The table below gives an idea of the preferred methods of disposal.

Weapon/Method		Male %	Female %
Gun		67	39
Knife		12	23
Beating		7.1	12
Other		7	12
Blunt Object		4.5	5.4
Strangle		0.7	0.9
Asphyxiation		0.6	2.6
Fire		0.46	1.5
Poison		0.4	2.5
Drowning		0.1	1
Explosives		0.03	0.07
Defenestration		0.02	0.04

By far, worldwide, the preferred method of murder by men or women is the gun[245]. I love the idea that we can still consider defenestration (chucking out the window) as a method of murder. There is something wonderfully old fashioned about it. Explosives lack a certain subtlety and I would imagine it would be quite difficult to obtain actual explosives nowadays, although you can buy pretty much anything on the net. Maybe there's a 'methodofdeath.com' which helps the would-be-murderer decide upon their favoured method. One of my previous students said that he had a syringe full of potassium at home ready for me if I mistreated him. Bless him. Teaching him English was enough to kill any normal man.

---

245 From 17,431 murders by women - Federal Bureau of Information Supplemental Homicide Report, 1999-2012. From 160,368 murders by men - Federal Bureau of Investigation Supplemental Homicide Report, 1999-2012. From: www.washingtonpost.com/news/wonk/wp/2015/05/07/poison-is-a-womans-weapon/ Dan Keating May 7, 2015

d) Surprise! A Bit About Serial Killers

And we all want to know about serial killers, don't we? Once you get your eyes hooked into the Sick Bastard UK TV channel at midnight, it's difficult to let go. There is something absolutely fascinating about someone who is willing to go to extremes to bump off apparently random people at apparently random places. It's very rarely random though, of course. There is, as the FBI have found, a commonality to serial killers, but that commonality is no less absorbing for all that.

First though, how do we define a serial killer? What is the actual definition of a serial killer? I shall pause here to let you get all those Corn Flake jokes out of your head and then move on.

Done?

Good.

There isn't a definition. Roughly, it is anyone who murders more than two people over an extended period of time. There is some difference of opinion over whether the number of kills should be above two or three. The definition below is about as close as I could find from the research.

> 'The different discussion groups at the Symposium agreed on a number of similar factors to be included in a definition. These included:
> • one or more offenders
> • two or more murdered victims
> • incidents should be occurring in separate events, at different times
> • the time period between murders separates serial murder from mass murder

> In combining the various ideas put forth at the Symposium, the following definition was crafted: 'Serial Murder: The unlawful killing of two or more victims by the same offender(s), in separate events'.[246]'

At any given time in the USA there are on average 35 serial killers. Since 1980 there have been (that we know of) 154 worldwide and 87 in the USA. At the moment, the FBI has 271 serial killer suspects on their lists[247].

Race (USA)	**% of Serial Killers (USA)**
White	52.21%
Black	40.3%
Hispanic	6.1%
Native American	0.8%
Asian	0.7%
Male	90.8%
Female	9.2%

It would be interesting to differentiate between male/female/black/white/Hispanic etc killers, but there seem to be few surprises in the list. 20.8% of these had military experience, but that means that 79.2% did not. Whether their military experience had any impact upon them becoming serial killers, I don't know, but I would imagine that it gave them an edge.

---

246 *Serial Murder: Multi-disciplinary Perspectives for Investigators*. Behavioural Analysis Unit, National Center for the Analysis of Violent Crime. Robert J. Morton Supervisory Special Agent Behavioral Analysis Unit-2 Federal Bureau of Investigation

247 *Serial Killer Statistics and Demographics*. www.statisticbrain.com/serial-killer-statistics-and-demographics/ 2016

Race (USA)	% of Victims (USA)
White	68.21%
Black	23.81%
Hispanic	6.53%
Asian	1.46%
Male	53.81%
Female	46.19%

The surprising thing, to me at least, about this list is that more men are victims of serial killers than women. Yet another false assumption on my part. Otherwise, it remains unsurprising except for the fact that Native Americans were not mentioned as victims. Whether this is because there simply weren't any, once again, I don't know.

The motives for murder by serial killers vary wildly, but the largest, claimed as enjoyment – including lust, thrill and power – is predictably high at 46.11%[244].

Motive (USA)	% (USA)
Enjoyment (thrill, lust, power)	46.11
Financial Gain	31.68
Multiple Motives	8.12
Anger	7.84
Gang Activity/ Criminal Enterprise	3.28
Avoid Arrest	1.21
Attention	0.62
Convenience	0.73
Hallucination	0.41
Cult	0.21

Presumably, the X-Factor has lowered the number of attention-seeking murderers along with all the other talent shows where the nutters put themselves on display.

Despite the above figures,

> '...it is important to look at trends across time," Aamodt told Business Insider via email. "For example, if you combine US serial killers across all decades, 52% of serial killers have been white [and] 40% black ... However, if you just look at the past three decades: 37% were white [and] 60% were black. A very different picture!"'[248]

Once again, there is context to take into account. If you broke these down into states, countries, even decades, the results might well present differently again.

> '...a greater proportion of the women, as compared with the general population, had: a history of having been physically or sexually abused; drug or alcohol problems; and a diagnosis or signs of mental illness...A striking contrast with male serial killers is the relative absence of sexual violence and deviance. Two exceptions were a female serial killer who was a rapist, and another who reportedly barked like a dog during sex. But overall, the researchers highlighted how the women in their study primarily killed for resources, while their male counterparts kill for sex. This follows evolutionary theory, Harrison and her co-authors explained, in the sense that men are said to be motivated more by seeking multiple sexual opportunities, while women are

---

[248] *Here's a Surprising Look at the Average Serial Killer* - Christina Sterbenz. www.businessinsider.com. May 2016

> motivated to find a committed partner with sufficient resources. The new analysis points to a worrying trend: a 150 per cent increase in the number of reported cases of female serial killers since 1975.[249]'

Neither is the 'typical' serial killer white, male and in their late twenties. In this age of equality and diversity, between 1990 and 2016, it's good to see that only 37.4% were white. And only 24.1 % in their mid-to-late twenties. White, male, mid-to-late twenties, that profile we all like to sit and smugly smile at when we have the killer at the end of *Midsomer Murders,* accounts for only 7.7% of the total[250]. However, the most popular method of death by a serial killer – strangulation - remains reliably high, followed by stabbing and shooting.

I suppose it is only sensible to expect the profile of killers to change over time. The world is constantly on the move, technology bounding ahead quicker than we can keep up and the influence of the media – the internet, TV, movies, YouTube, social media, newspapers, books – greater than it has ever been, all of it trying to find a way into our psyche. How much damage it all wreaks I'm not sure, but it is not possible to constantly be bombarded by the same messages without being affected – isn't that brainwashing, after all?

Finally, if we have to make this into a competition (we do, we do), England gets silver in the serial killer medals tables[251]. I'm not sure that this is something we should shout about, but with football, cricket and just about every other thing that we try on the world stage ending in dismal failure,

---

249 The Psychology of Female Serial Killers - Research Digest. March 5, 2015. Post written by Christian Jarrett

250 Aamodt, M. G. (2016, September 4). *Serial Killer Statistics.*

251 Source: Radford/FGCU Serial Killer Database Updated: 9/4/2016

it's nice to know that we can at least make a showing in the lunatic fringe stakes.

SERIAL KILLERS BY COUNTRY			
COUNTRY	NO. OF SERIAL KILLERS	% OF SERIAL KILLERS	SHARE OF WORLD POPULATION (%)
UNITED STATES	3204	67.58	4.35
ENGLAND	166	3.5	0.71
SOUTH AFRICA	117	2.47	0.74
CANADA	106	2.24	0.49
ITALY	97	2.05	0.80
JAPAN	96	2.02	1.70
GERMANY	85	1.79	1.08
AUSTRALIA	81	1.71	0.33
INDIA	80	1.69	17.81
RUSSIA	73	1.54	1.93
FRANCE	71	1.50	0.87
CHINA	57	1.20	18.56
MEXICO	37	0.78	1.73
BRAZIL	27	0.57	2.81
AUSTRIA	22	0.46	0.12
HUNGARY	20	0.42	0.13
SPAIN	17	0.36	0.62
POLAND	15	0.32	0.52
SCOTLAND	15	0.32	0.07
NETHERLANDS	12	0.25	0.23
SWEDEN	12	0.25	0.13

It is fascinating to look at the figures. The supposed modern cradle of democracy, the USA, has more nuts than a Snickers bar. It has 67.58% of the worlds serial killers for only 4.25% of the world population. Sweden's share of the world population is 33 times smaller, yet they have a serial killer rate which is a staggering 267 times less than the USA. Why? Neither of them belongs in the 'third world' category. Neither of them is poor. Sweden's population is a few thousand over 10,000,000; America's is about 329,000,000. About one-fifth of Sweden's population has an immigrant background. It has a life-satisfaction score of 9.1/10. I only rate my own life-satisfaction score at about six. Without wine and pork scratchings it would be a four-at most. America, as we have already seen is an enormously diverse place. Is diversity and ethnic tension at the root of the psychology that leads to a nation producing serial killers? Religious extremism? Extremes of poverty and wealth? Simply a large population? Nature or nurture? Sweden has always been looked upon as a more enlightened society, one that encourages equality, whereas America, far from being the modern cradle of democracy, is the cradle of capitalism, of the 'individual first' philosophy, is a modern empire-builder that builds not on the physical conquering of lands, but the financial and psychological capture of populations by use of media and financial bullying and manipulation; with financial power comes physical power and that option is always left to lie in the background like a badly trained Rottweiler.

China's share of the world population is 4.2 times greater than the USA and yet it has 56 times fewer serial killers. It could be that China has a less stringent practice of recording of such things and a reluctance to share such information on the world stage; it is a secretive and oppressive dictatorship which is very sensitive about what is shares with the world, after all.

The discrepancies between countries are too numerous to list. One has to take into account not only diversity in skin colour, but in religious belief and in culture, in wealth and in poverty, in mental and physical health, in styles of government and even in the climate. Every aspect of our daily lives dictates what we are and what we become.

There is a whole lot more about serial killers out there – yes, they do often have a history of animal torture and bedwetting at a late age and family life was probably unstable - and it changes day by day. The same applies to the gender differences. One would hope that as we become more enlightened, the abuses against women would drop, but at the moment, the gap between rich and poor in this world is as bigger as it has ever been and the methods of killing each other more numerous as technology surges ever onwards. Until the human race reaches its own version of homeostasis and I have no idea what that state is, then I suspect that the problems will be ongoing.

PART 6

The Whole Death Thing

Just Leaving…

There are many, many reasons why I shouldn't have been a nurse. For one, I'm not a people person, which in a job like that is a distinct disadvantage. I don't like petty politics either, which in a system that thrives on petty politics in the way that bacteria thrive on the faecal debris in the gut, is also a hindrance. Another reason is that I'm not a fan of death. In fact, I would go so far as to say that I am 100 % against it. The problem was that I didn't realise this until it was much, much too late. With the combination of politics, people and death, sat like a bunch of barrels of gunpowder under Parliament just waiting for a spark, it was just a matter of time. Then, one day, someone walked in with a box of matches and blew the shit out of everything.

It wasn't until I was able to view my time in nursing in retrospect that I realised that I didn't have the temperament or, frankly, the will, to deal with it. You either have to be stupid to remain unaffected or very self-aware. I was neither, although that is open to debate. There is a happy medium which many people find, my wife included, where they can balance that tightrope between the personal effect it has upon them and the subsummation of their better selves by the constant barrage of need and despair. There came upon me a realisation, brewed over many years and now strong enough to pour, that none of us get out alive and that, no matter how good, how bad, how fit, how careful, how clever, how thick, how fat, thin, tall or small we are, there is only one inevitable outcome.

So, we all die. Some of us are lucky enough to die quickly, bang, either in an accident or by an elephant-sized heart attack. I think it is much easier to greet St Peter with the words, 'What the fuck just happened?' than with, 'Bloody hell! That took a long time'.

Accidents and slow deaths do of course have one big thing in common: you die. The actual physicality of the process, *post-mortem*, remains the same. It was the people who lingered that I felt sorry for, who had time to think about it, who had to deal with the blunt hammer of the psychological trauma as well as the act of death.

I remember, when I worked on a surgical ward in Barnsley Hospital, there was a chap who was in a side room on another ward. He had, if I remember correctly, started out on our ward and then moved across the corridor to a sister ward, in order to create bed spaces. We had remained on nodding terms.

I walked past his room one day, running an errand between wards and he beckoned me, asked if I had a minute. Of course, I said. He was my father's age, had that same air of respectability that my father had, and so I think I had an affinity for him.

He told me that the news had come through that he was dying. He was still at the rabbit in headlights stage, but the logical part of him, that part which needed to find a coping mechanism and be a step ahead, wanted to know what happened when he died. He asked me, outright. I told him. He thanked me and I left.

I never saw him again.

This part of the section is going to deal matter-of-factly with death. It's not about the psychology, only about the way we shut down, die and then decompose. Elements of the psychology of the subject might well slip in, but that is not primarily what it is about. That comes later.

The Way to Dusty Death

There are two types of death:

1) Clinical Death:

   In a state where vital signs – such as respiration, heartbeat and corneal reflex – are not present, but from which patients are sometimes revived.

2) Biological Death:

   The death caused by degeneration of tissues in the

brain and other parts. Most organs die after biological death.

In the case of organ transplants (a spritely little conversation has grown around this subject in our living room), www.organtransplants.org says:

> 'Organ and Tissue Donation after Cardiac Death
>
> Typically, when a person suffers a cardiac death, the heart stops beating. The vital organs quickly become unusable for transplantation. But their tissues – such as bone, skin, heart valves and corneas – can be donated within the first 24 hours of death.'

This is further clarified at cprtraining.blogspot.co.uk:

> 'Clinical Death and Biological Death
>
> When the heart stops pumping the lungs stop breathing, the person is in a state of clinical death. Usually, this is measured from the moment the cessation of heart and lung functions begin.
>
> Most often the timetable is 0 to 4 minutes. In this instance, remaining oxygen in the blood can still supply the brain with enough oxygen without brain damage. However, after 4 minutes, the oxygen remaining in the blood will be used up and there is already the possibility that brain damage will occur. At this point in time, the person is said to be in the state of biological death.
>
> To summarize:

> CLINICAL DEATH (0-4 minutes) - occurs from the time the person stops breathing and the heart stops pumping.
>
> BIOLOGICAL DEATH (4 minutes onwards) - brain damage may occur at this point.[252]'

Biological death cannot occur without clinical death.

In the future, there is the possibility that non-heart beating organ donations might well be used. This could be because the demand for donors is now so high and because targets are not being met that the goalposts need to be changed (cynical view) or that research is finding that the critical time between clinical and biological death is becoming less relevant with the advances in technology (the less cynical view). Research is ongoing.

So, what is the process that leads to biological death? In the case of those who do not die instantly by accidental means, there is a process that is dictated by both the psychology and the physiology of the individual and also the environment that they are in. This is one of those areas where it is almost impossible to separate psychology and biology because they have a massive impact upon each other.

In the time leading to death, which might be several days or just a few hours, the body goes through certain changes common to us all. The closer to death the individual is, the more apparent the symptoms become.

> Sleepiness and difficulty waking (semi-conscious)
> Difficulty swallowing, loss of ability to swallow or not wanting to eat or drink at all
> Loss of control of bladder and bowel control
> Restless movements (as though they are in pain)
> Changes in breathing pattern

---

[252] http://cprtraining.blogspot.co.uk/2008/05/clinical-death-and-biological-death.html

Noisy breathing
Cold feet, hands, legs and arms/cooling
Reduced intake of food and fluid
Confusion and disorientation[253]

Due to the disease process, the body is shutting down. The normal functions of the organs are affected so, for example, the liver, if cancerous or affected by disease elsewhere, can no longer carry out its many functions.

The human body is made up of eleven organ systems that work with one another. These systems include the integumentary (skin, hair) system, skeletal system, muscular system, lymphatic system, respiratory system, digestive system, nervous system, endocrine system, cardiovascular system, urinary system, and reproductive systems[254]. When one of these goes haywire, it has a knock-on effect – when one organ stops functioning properly, this will inevitably impede the functioning of one or more of the other organs in the body. If the lungs, for example, fail to take in oxygen due to emphysema, this affects the distribution of oxygen to the muscles, which leaves the individual unable to perform tasks of any duration. The body fails to maintain what is called **homeostasis**, which is a stable balance between all the elements within the body. Right now, I hope, you are in homeostasis, that your kidneys are functioning well, which allows your blood to be free of toxins and that your liver is functioning well, which avoids hypertension and all the other nasties that we mentioned earlier.

As disease progresses, the body can no longer synthesise the food it takes in. Neither might the individual actually want or need to eat. There is the psychological 'what is the point?' aspect to this which might cause them to reject food.

The individual, due both to the disease process - they might have difficulty swallowing, might be vomiting and unable to keep anything down - and the lack of diet, is essentially running

---

253 www.cancerresearchuk.org and www.commtechlab.msu.edu and http://health.howstuffworks.com

254 http://anatomyandphysiologyi.com/human-body-organ-systems-an-orientation/

out of energy. The lack of energy leads to an increased need by the body to rest and preserve the little energy it has. This sleepiness will eventually become loss of consciousness. It might well be accelerated by the use of opiates in pain control.

With the lowered intake of diet and fluids, the body slides into a chemical imbalance. This chemical imbalance often leads to confusion. Physically, if the confusion is not due to the disease process itself, e.g. primary or secondary cancer in the brain, then the confusion might well be due to anoxia (lack of oxygen) or dehydration and /or any urine infection that follows this. A urine infection, untreated, as it probably would be in the very terminally ill, would possibly lead to systemic infection.

The body eventually begins to relinquish control and this might lead to urinary and faecal incontinence. These in turn might promote the formation of skin irritation and/or pressure sores which, as well as being distressing to the individual, can lead to septicaemia through infection of the affected areas.

The lack of mobility allows fluids, normally cleared by coughing and possibly increased by heart failure, to pool and impinge upon breathing. This can lead to a chest infection and, once again, septicaemia and death or to anoxia and death.

Breathing can also become very loud due to the fluids pooling in the lungs and in the trachea. Eventually this can turn into something called Cheyne-Stokes breathing, also known, when accompanied by loud, moist breathing, as the 'death-rattle'. At this stage, the body is on automatic. Breathing is normally stimulated by the amount of carbon dioxide in the body and is controlled within the brain. Respirations can become horribly slow to the onlooker, sometimes as low as one breath a minute. The heartbeat might actually increase in an attempt to compensate for the lack of oxygen and push the oxygen in the blood around the body more quickly. The pulse will often become irregular and gradually slower.

The high percentage, at about 12%[255], of restless and

---

[255] Lichter, I. and E. Hunt. *The Last 48 Hours of Life.* Journal of Palliative Care 1990.

agitated patients, suggests that altered states are not uncommon in dying patients. The difficulty for those observing the process is to differentiate between pain and restlessness/confusion due to anoxia and/or infection. We all move in our sleep; dying people are no different. Just because you are dying does not mean that you don't dream or react to your environment.

Depending upon the environment and the state of dress of the individual, the body will begin to cool, the speed of cooling dependent upon these factors. This is because there is a slowing of the circulation and a poor perfusion of blood to the extremities. The individual will lose colour. Their extremities such as fingers, hands, toes and feet, even nose and lips and the tips of their ears, will sometimes become purple as the blood no longer reaches those peripheries and they begin to die.

> 'At the end of life, the chemical balance of the body becomes completely upset. The dying person then slips into unconsciousness. This is usually right towards the end, maybe only a few hours or days before death. Breathing becomes irregular and may become noisy. You won't be able to wake them at all. Their breathing will stay irregular for some time and will stop at some point.[256]'

> 'Clinical death occurs when the person's heartbeat, breathing and circulation stop. Four to six minutes later, biological death occurs. That's when brain cells begin to die from lack of oxygen, and resuscitation is impossible.[257]'

Out, out, brief candle!

---

256 www.cancerresearchuk.org/about-cancer/coping-with-cancer/dying/what-happens-in-the-final-days-of-life#gBxMa924V5Elv1bi.99

257 http://health.howstuffworks.com/diseases-conditions/death-dying/dying-process.htm

Assuming that the person is not for resuscitation (in a controlled environment), that help is not at hand or does not simply recover on their own, we now begin to cross from the area of clinical death into biological death. There is, depending on who you read, a time of about four to six minutes until brain and tissue death occurs and such things as organ donation and recovery are no longer viable options.

When a person dies:

- The person cannot be woken up
- The heart stops beating
- Breathing stops
- Body color becomes pale
- The body cools
- Muscles relax
- Urine and stool may be released
- The eyes may remain open
- The jaw can fall open
- The trickling of internal fluids may be heard[258]

A doctor can diagnose a patient dead when

- there are no palpable pulses, indicating that the heart has stopped.
- there are no heart sounds on auscultation (listening) (or asystole – shown by a flat line indicating absence of heartbeat - on ECG).
- there is no observed respiratory effort, indicating that no gaseous exchange is taking place and that vital organs are no longer being oxygenated.
- there are no breath sounds on auscultation.

---

[258] www.commtechlab.msu.edu

- the pupils are dilated and not reactive to light[259], indicating no brain activity and a relaxation of the muscles in the eye. They constrict again when *rigor mortis* sets in. After up to a few hours, the eyes become cloudy due to a release of potassium from the breakdown of red blood cells.

The individual should be observed by the person responsible for confirming death, for a minimum of five minutes to establish that irreversible cardiorespiratory arrest has occurred[260].

'English law

•Does not require a doctor to confirm death has occurred or that "life is extinct".
•Does not require a doctor to view the body of a deceased person.
•Does not require a doctor to report the fact that death has occurred.
•Does require the doctor who attended the deceased during the last illness to issue a certificate detailing the cause of death (unless the death is referred to a coroner).
So, a doctor's legal duty is to notify the cause of death, not the fact that death has taken place. Doctors, nurses or suitably trained ambulance clinicians may confirm that death has taken place. There is no legal obligation on a doctor to see or examine the deceased before signing a death

---

[259] Oxford Handbook of Clinical Surgery. GR McLatchie. 1990 and http://patient.info/doctor/death-recognition-and-certification
[260] http://patient.info/doctor/death-recognition-and-certification

certificate. This is the case across the UK[214 261].'

I did not know this – until now, obviously.

When the heart stops beating the body will begin to lose temperature. The body usually maintains a temp of about 37°C. This is called **Algor Mortis** (Latin: Algor = coldness. Mortis = of death). The body's temp will drop by about 1.5°/hour until it reaches the same temp as its environment.

The other thing that happens, after twenty to thirty minutes, is something called **lividity** (Latin: *livor mortis* – Livor = bluish colour. Mortis = of death. Hence 'livid' for 'angry'). All the fluids in the body basically follow gravity and pool at the lowest point. Usually, the patient is on their back, so the effect will be of a very pale upper half(ish) to the flat body with a very blue/purple hue to the lower part of the body, where blood and other fluids have gathered[262].

It is also at this stage, within just a few minutes of death, that **autolysis** begins. This word comes from Greek, which can be broken down as Auto = self and Lysis = splitting or digesting. The lack of oxygen in the body means that the actual cells themselves cannot get oxygen and they become acidic. Enzymes, tiny molecular catalysts which are there to cause chemical reactions in both life and death, start to eat the cells and the first to go are usually the liver, which is rich in enzymes and the brain. Bacteria within the gut, at the same time, will begin to digest the soft tissues.

Scientists are now getting all hot under the collar about the microbes in our bodies and the effect they have within a sort of 'ecosystem' which is called The Gut Microbiome.

> 'A microbiota is defined as the community of microorganisms, including bacteria, archaea…

---

[261] *Confirmation And Certification Of Death;* British Medical Association (BMA) guidance for GP practices

[262] Image: By goga312.Goga312 at ru.wikipedia from Wikimedia Commons

> viruses living in a specific environment. A microbiome, on the other hand, is the entire collection of all the genomic elements of a specific microbiota... we know that the gut microbiome plays a role in functions required for the physiology and correct development of our organs, and that its composition is related to aging, environmental factors (diet, physical activity, etc.), and pathological conditions. Understanding the role of specific microbes will open the way to novel strategies for disease diagnosis, monitoring and therapy.[263]'

Not only are there now implications regarding the bacteria with the metabolic and developmental processes of the body, but also with regard to the disease process when we are alive, linking it to such things as obesity, inflammatory bowel disease and autism.

There is also the theory that these microbes target different areas of our body in a particular order at a particular *post-mortem* time, for example reaching the liver at about twenty hours after death.

> 'Most internal organs are devoid of microbes when we are alive. Soon after death, however, the immune system stops working, leaving them to spread throughout the body freely. This usually begins in the gut, at the junction between the small and large intestines. Left unchecked, our gut bacteria begin to digest the intestines – and then the surrounding tissues – from the inside out, using the chemical cocktail that leaks out of damaged cells as a food source. Then they invade the capillaries of

263 *The Role of the Gut Microbiome in the Healthy Adult Status.* Valeria D'Argenioa, Francesco Salvatore. December 2015. www.sciencedirect.com

> the digestive system and lymph nodes, spreading first to the liver and spleen, then into the heart and brain…Javan's study suggests that this 'microbial clock' may be ticking within the decomposing human body, too... Thus, after we die, our bacteria may spread through the body in a systematic way, and the timing with which they infiltrate first one internal organ and then another may provide a new way of estimating the amount of time that has elapsed since death.
> "After death the composition of the bacteria changes," says Javan. "They move into the heart, the brain and then the reproductive organs last." [264] [265]'

Rigor Mortis[266], that favourite of Quincy MD (get with it kids; he was a dude, a shouty, go-out-on-a-limb kind of dude) begins to set in at between 2 to 6 hours and 'typically lasts for 24-48[267] hours[268]'.

The reason that *Rigor Mortis* occurs is because the muscles, actually the cells within the muscles, contract, starting with the head and neck, then the torso and then the limbs. The contractions are caused by two proteins, actin and myosin, long, thin, and stringy proteins which make up microfilaments, both of which would normally help in the smooth working of the muscles. Myosin and actin work together to help the muscle cells relax and contract. The two proteins need each other and together they are

---

[264] Gulnaz Javan et al. of Alabama State University in Montgomery. http://gizmodo.com. August 2014

[265] If you want to know more about the Microbiome, there's a good explanation, understandable by idiots like me, at http://www.slideshare.net/Mahmoud-Shab/human-microbiome-44681820. It's an interesting read.

[266] Yep, it's from the Latin - Latin: rigor "stiffness", mortis "of death".

[267] 72 hours according to Anne Marie Helmenstine, Ph.D. on chemistry.about.com

[268] http://aboutforensics.co.uk/decomposition/

called actomyosin. Combine those protein threads with some ions in the muscle cell and you get a huge contraction[269]. Because calcium, which is normally kept outside of the cells and helps the proteins function, is able to get into the cells, this stops the two proteins working, essentially freezes them; adenosine triphosphate (ATP), which stops the action of calcium, is no longer working due to the body's lack of oxygen[270] and the muscles contract to their fullest, resulting in some of the strange postures that are seen *post-mortem*.

It is also at this stage that flies, blowflies and houseflies, begin to take an interest in the meal/breeding ground laid out before them. They lay eggs in any opening in the body, and I mean *any*, within about twenty-four to thirty-six hours of death. They lay their eggs, which hatch within about twenty-four hours and then start nibbling after burrowing into the body. They will become fully fledged flies after two to three weeks[271].

Think about that next time one lands next to your sandwich.

Life's but a walking shadow

There are five stages to the decomposition of the body:

1) Fresh – which lasts one to two days (days one to two)
2) Bloated – which lasts about two to six days
3) Decay – which lasts about five to eleven days
4) Post-Decay – which lasts between ten and 24 days
5) Dry Stage/Skeletal Stage – twenty-four plus days.

Between the second and the eleventh days (stages two and three), the body begins to putrefy. It is basically melting into soupiness. The anaerobic bacteria (ones that don't need oxygen to survive), that live in the gut, set about their orgy of feasting and in

---

269 www.biology4kids.com/files/cell_microfilament.html and Image
270 www.yalescientific.org/2010/02/everyday-qa-what-causes-rigor-mortis/
271 http://australianmuseum.net.au/image/initial-decay-0-to-3-days

doing so produce copious gases, which leads to bloat. These gases might well lead to a rupture of the inner organs and even perhaps of the skin, usually between four and ten days[272]. This is an open and often irresistible invitation to any passing scavengers. A week after death, the skin has blistered and the slightest touch could cause it to fall off[273].

With the process of putrefaction, the skin begins to change colour and becomes greenish.

> 'As the body continues to putrefy, the skin blisters, hair falls out and the fingernails of the deceased began to sink back into the fingers. These skin blisters are also filled with large amounts of liquid just as in a blister you might get from running or walking too far. The body's skin tone then becomes what is known as 'marbled'; an intricate pattern of blood vessels in the face, abdomen, chest and other extremities becomes visible. This is the result of the body's red blood vessels breaking down, which in turn release haemoglobin. As the process reaches its conclusion, the body will now be almost black-green and the fluids – known as purge fluid – will drain from the corpse. This happens normally from the mouth and nose but can also occur from other orifices. The body's tissues then begin to break open and will release gas and other fluids in the same way as a fruit that has been left too long in the sun… It is also important to note that the internal organs of the deceased will begin to decay in a particular order; beginning with the intestines, which as well as holding bacteria also hold various levels of acidic fluid which – when unable to circulate – begin to eat through their surrounding

---

272 http://education.seattlepi.com/stages-human-decomposition-process-4705.html

273 http://health.howstuffworks.com/diseases-conditions/death-dying/dying4.htm

tissues. As the intestinal organs decay so too do the liver, kidneys, lungs and brain. [274]'

At ten to twenty-four days, what is left are the harder tissues such as cartilage, hair, bones and the wet, sticky remnants of the second parts of the process. Insects that prefer a drier environment are present such as beetles and cheese flies. These larvae are able to chew through the tougher materials that remain such as ligaments and cartilage.

The final stage, the dry or skeletal stage, is when the wetness is gone. The dry tissues are slowly removed by remaining insects such as beetles and flies, while moth and larvae finish off the hair.

The death of an individual, whether it be a bird, a dog, a blue whale or a human, produces in itself another ecosystem; life itself flourishes from death.

Ashes to Ashes, Dust to Dust – The Disposal of the Dead

How much does it cost to die? (it's free actually; it's the burying that's the bugger)

I'm in the wrong business. Actually, being unemployed at this moment in time, I suppose, strictly speaking, I'm not in any business, but if I were, I'm pretty sure I would be in the wrong one.

It would seem that the best place to be would be, and forgive me for this, in a dying business. Funerals. Do you remember in those old cowboy films how they would quickly kick some dust over a corpse, jump back onto their horses and ride into the sunset? Nowadays that would cost you anything from £900 to £8000[275]. Still, that's cowboys for you.

---

[274] www.exploreforensics.co.uk/the-rate-of-decay-in-a-corpse.htmlDecay (5-11 days)

[275] www.ft.com. April 12, 2016.Paul McClean

We are in a bit of a Catch 22, are we not? We don't want Cousin Wally turning to soup on the sofa or the health risks inherent with the long-term storage of a loved one. Not only that, but the law would not allow it. One way or another, Wally has to go.

The funeral market is worth about £1.7 billion a year[276]. That is a lot of money. It is probably the only business that hasn't suffered in the recession, although there is no doubt that customers have cut back on those little extras that they might have paid for prior to losing their job.

The average cost of a funeral varies depending upon whom you read. The BBC[277] (Gawd bless 'er) says that the average cremation costs £683, while the average burial costs £1,645. That however is for the basics. That gets you four cowboys with dusty boots and cold saddles. Extras, they say, could bring the cost up to £6,000.

Another estimate is £2,801 as the average cost with the lowest being £1,995[278] (why not £1995.95 to really press home the cheap and nasty psychology. There is also something satisfying about getting 5p change), but once again this excludes third party costs. For this, you get, and I quote:

- 'We'll remove your loved one from their place of death into our care
- If death occurs out of normal working hours, at home or in a care home, we'll remove your loved one into our care, out of hours, at no extra cost
- We'll prepare, care for and dress your loved one in a gown to match the interior of our 'Simple Coffin'
- Your loved one will rest in our mortuary facility, or if required, a private rest room

[276] *Funeral Business Reaps Profits As UK Death Rate Soars*. www.ft.com. April 2016.
[277] www.bbc.co.uk. Paul Grant. 4 October 2015
[278] www.co-operativefuneralcare.co.uk

in our funeral home

- You'll be able to visit your loved one during normal working hours at no additional charge
- We'll transport your loved one from the funeral home to the crematorium or cemetery in a hearse at the appointed time of the service
- We'll make all necessary arrangements for a dignified funeral
- Your Funeral Director will be there at the funeral to support and guide you through the funeral.'

This however does not include those mischievous 'third-party' fees such as 'fees for the doctor, minister, church and crematorium (and cowboys).' They go on to say that 'On average last year (2015) our clients paid £1,016 in third party fees.'

Third party costs include:

Cremation fees
Doctors' fees (Not applicable in Scotland)
Officiant/minister/celebrant fee to lead the service

For burial third party fees usually include:

Interment (burial) fee
Service fee (at a church or other venue for example)
Officiant/minister/celebrant fee to lead the service
The cost of the coffin or casket
Professional services (including liaisons with third parties)
Funeral home facilities (for example the Chapel of Rest)
Presentation of your loved one
Hearse/Limousines
Administration charges

Staff for the funeral itself (including bearers)[279]

So how much are these third-party costs? Well, for a start 'before cremation can take place, two certificates need to be issued – the Certificate of Medical Attendant (Form 4) and the Conformatory Medical Certificate (Form 5). Each form will need to be certified by a different doctor at a cost of £82 apiece'.

These are forms to confirm that the burnee is actually dead and that there was no foul play. Really? So, the first doctor a) was less capable of saying whether someone is dead and b) is not either able or permitted to have suspicions? The death certificate costs £9.25[280] (at the moment) and then you have to pay for extra copies at, for all I know, £1,000,000 each, and that's without the gold-leaf edging.

Mark Mason from *The Spectator* wrote an article in May 2015 called *How Your Funeral Director Is Ripping You Off*. His partner's father had died and he was good enough to itemise the costs:

1 Bog-standard Cremation (his words, not mine) - £4000
Flowers - £150
Urn - £66
Crematorium costs - £852
Service Stationery - £76 ('20 copies of the words to the hymn we'd be singing')
Doctor - £160
Coffin, cars and staff - £2,865

How do you justify these prices? £150 for flowers? I would expect orchids from the very peaks of the Himalayas flown in by the ghost of Edmund Hilary and carried by Sherpa Tensing for that money. The urn? Well, it was apparently aluminium. That price is probably about right after a quick online search, but I would think that this is something that Wilkos and Poundland

[279] www.co-operativefuneralcare.co.uk and www.funeralzone.co.uk
[280] www.funeralzone.co.uk

could look into. The cost of a coffin? Another quick search reveals anything from £155 up. You could hire a van from Darton Van Hire (a Barnsley firm; others are available nationally) and get some mates to carry the coffin for the cost of a pint each.

I exaggerate, but you see where I am going with this?

It does tick me off that, when people are at their most vulnerable, their most defenceless, the vultures drop down and start picking bits from the cooling corpse.

I've told the wife, buy some lighter fluid and a box of matches, stick some spuds for baking under me and make an evening of it.

The two main funeral service companies in the UK are Dignity and the Co-op. They are doing very well. Shares in Dignity went up by 311% between 2010 and 2015 (the Co-op are not listed). Dignity's pre-tax profits for 2015 had grown by almost a quarter[276]. Luckily for them, dementia is on the march and death rates rose at their fastest pace since 1968. Their revenue for 2015 was £305.3 million. 'Dignity said underlying pre-tax profit rose 23 per cent, to £72.2m, on sales that were 14 per cent higher, at £305.3m.[274]' In the Co-operative's 2015 Annual report, their revenue was £399 million with an underlying operating profit of £78 million, an increase of 18.2% in profits.

Even in death, it seems that the privileged get a better send off and the mourners derive more comfort with less penny-pinching guilt.

I know, I'm probably being over-sensitive in the 'us and them' stakes but, from birth to death, it would appear that if you have the money, you will have better things and achieve better things, even in death. Am I envious? Maybe, yes. Am I annoyed by the disparity between the haves and the have-nots? Assuredly.

The Laws of Deathliness

I am frankly surprised at how lax the laws are when it comes to doing away with Cousin Wally. It seems that you really could get away with stuffing him into an old suitcase and popping him on top of the wardrobe along with the dust-bunnies, the glossy magazines and those long-legged, small-bodied spiders. He would, of course, begin to honk something terrible after a while, but if you can put up with that until the dry stage, well, there's up to £8,000 in your pocket minus the cost of a suitcase. The only third-party involved here is Luis Vuitton or the suitcase manufacturer of your choice.

However, it is clear that someone has to take responsibility for Wally. If it's not his immediate family, then the state will have to take responsibility for him.

And, no, the top of the wardrobe was not really an option.

Nobody, it seems, has the right to a body; you cannot own it, possess it or in any way claim it as yours. Unless you have been nominated. Lucky you. This is where the word 'duty', a useful little legal word, comes into play.

The law is quite strict about what you can't do with a body. You cannot:

Detain a body
Refuse to deliver it to the executors for burial
Conspire to prevent a lawful and decent burial
Dispose of the body to prevent an inquest
Sell the body for dissection
Expose the body in a public place if to do so would shock public decency[281].

I believe it would be difficult to shock public decency nowadays, even with a naked, pornographically *rigor mortis* body. I think people would be more outraged if I started shouting mildly sexist slogans than if I were dragging dead Wally through town.

Going back to who owns the body – it is basically the executor of the will (this is the moment when you regret saying 'Yeah. Sure, buddy. No problem!', just to change the subject that drunken night in the pub), the parents or 'the householder of where the body lies'. How is that for a money-saving thought? Dump Wally on the front doorstep of that neighbour you hate and

[281] www.contactlaw.co.uk

they are then obliged to bury him. And get this: 'It is an offence for any such person, having sufficient means, to fail to discharge this duty'.

Screw you, neighbour! Have a dead Wally!

If it's any compensation, you don't have to bury Wally in a cemetery or any other 'authorised' place. You can bury him in your garden if you wish, so long as it is private ground and provided that you don't cause a nuisance. It's almost worth finding a body just to see where the neighbours draw that 'nuisance line'. Where my parents live, they can't hang washing out on Sundays. Seriously. What the hell would a burial do to those weekend-dirty loons?

You can't burn Wally in your garden either, that's just crossing the line; you can only burn people in crematoriums, presumably because nobody wants some sort of nightmarish Dickensian[282] image of Wally's fatty globules sliding down the neighbours freshly laundered sheets as his melted remnants are carried over the fence on the merest breeze.

You can pretty much scatter ashes anywhere, except where permission is needed but, to be honest, if they say no, all you do is use the *Great Escape* method and dribble them gradually out through a hole in your trouser pockets[283]. You can't *own* ashes either, apparently. They still have the same 'rights' as a body; better left to drift upon the four winds or hopefully towards warmer climes upon the oceanic tides, I think. That would suit me, though I'm willing to bet I'd end up in that sink-hole of waste in the Pacific, the Great Pacific Garbage Patch. It's real. Google it.

That Ole Black Magic…

The thing that amazes me about religion is not what the differences are, but what they all have in common.

They have a god, obviously. They all believe that the

---

282 Charles Dickens' *Bleak House*. Mr Krook's grisly demise is always worth a reread.

283 How many more times? Watch the bloody film! (You have to say all that in a Michael Caine voice)

actions of this life affect what happens in the afterlife. There is a Day of Judgment in both Islam and Christianity and they all have holy books. Neither can you pray to any other gods. There is only one, very proprietorial, Supreme Being. They all have their version of the seven sins somewhere in there and they all have priests in some form and it is usually only through these priests that you can truly access your god.

They also all rely upon the gullibility and fears of the population. This is not a statement of bias. It is a statement of annoyance because state control, financial control and hierarchical control within the community have all at some stage been driven by religion. Indeed, they still are. In the west we now see the extremes of religions such as Islam to be somewhat fanatical but the truth is that it wasn't so long ago that religion had the same influence upon our lives. Edward I, Richard I, Henry VIII, Charles I and Mary all used religion to broker power (actually all the Royal Houses did up to the mid/late seventeenth century and even after that it had an influence as to who sat in the chair; I'm just picking on them because they come to mind), not all successfully, but the effects of their meddling in religion inevitably led to deep and long-lasting changes in society and the consequences of those changes, for good or bad; perhaps even the death of religion as a controlling medium in the west.

Those are some of the bad points. The good point of religion is that it is a coping mechanism, a belief structure that helps the individuals within society, and therefore society itself, to cope in times of stress and death. In the context of this book (the context of which, I admit, has occasionally wandered), it is very much one of the psychological coping methods common to all societies and has been since man first looked at the moon. It has been used to justify the imposition of death, during the crusades for example or in the modern suicide bombers or the Islamic Front war, and it is used to combat death, to make death bearable, to give comfort to those on their way and to those inevitably left behind.

We have managed to construct elaborate concepts and processes through which we must go to purify our souls, which those still living must adhere to in order for us to safely reach the other side. The rules we construct go against those animal instincts that thrive so strongly within us while we are alive, perhaps to compensate for the times that we have behaved with instinctive baseness.

It is also a part of the mourning process, those stages that we all go through, whether as the victim or the onlooker, which enable us, most of the time, to get through a traumatic experience.

The human psyche is a very fragile thing and we have managed to build this construction of words and deeds beneath which we can shelter while the storm passes. If we do not use this shelter, then we find it difficult to bounce back and sometimes never do. We do it all the time, at work, in the home, at play, in our social circles, but with the finality of death, with such a large leap from the known to unknown, the coping mechanisms have to be that much greater. We have built industries upon these insecurities – funeral parlours and the established church are no more than industries looking to establish a profit. Should it be 'Thou shalt have no other gods before me'? Or 'Thou shalt profit no other gods before me'?

I'm saying all this and I consider myself a Christian. There has always been a conflict within me as to whether I should commit myself wholeheartedly, but then that old question arises which blocks the essence of pure religion: unquestioning faith. I envy those who have it and at the same time feel a certain freedom in my unwillingness to submit to myth and swindlers without question.

BUDDHISM

Buddhists believe in reincarnation. The ultimate goal is to hop off this constant cycle of birth and death and attain nirvana, that is, complete enlightenment. Buddhists believe that death of the physical body is certain, that the body is simply a vessel used to gain the enlightenment needed to attain nirvana. Cremation is the preferred mode of disposal of the body.

CATHOLICISM

Hitler could go to Heaven. If, as the bullet cracked open his skull in those final milliseconds in the bunker, he repented and sought forgiveness, he could have gone up, not down. In theory, Fred West could also go to Heaven, as could Harold Shipman, the Yorkshire Ripper, George W Bush and the feller who shot Lincoln.

The Catholic Church is very ritualistic. It is argued that it maintains this ritualism, and the elitism that goes with it, in order to keep the ordinary man in his place and maintain the mystery and power of the established church. This is why the church was so reluctant to have the bible translated from Latin into a common language – it took away that elite knowledge to which only the educated had access. The idea that one man, the priest, should have upon his tongue the power to send you to Heaven or leave you to Hell is a very empowering, and disempowering, concept.

Catholics also have Purgatory which, contrary to popular belief is not some sort of waiting room, but a place where the final purification of the soul takes place in order to obtain the holiness necessary to enter Heaven.

According to Catholic belief, the bodies of the dead will be resurrected at the end of time.

> 'The Catholic funeral service is called the Mass of the Resurrection. During it, Jesus Christ's life is remembered and related to that of the deceased. Eulogies are not allowed during the funeral mass, but may be delivered at a wake or other non-religious ceremony. There is also a final graveside farewell, and additional traditions depending on the region. The Church encourages Catholics to be buried in Catholic cemeteries.[284]'

It wasn't until 1963 that the Catholic Church allowed

---

[284] www.beliefnet.com

cremations. The church does stipulate however that the remains must be interred, not left on the mantelpiece.

When a person is dying, the priest can deliver what is known as the Last Rites. During this, the dying person is anointed with holy oil.

> 'The Church developed the ritual of last rites to prepare the soul of the dying person for death and for the individual judgment to come. That is why confession of one's sins, if the dying person is able to speak, is an essential part of last rites; having confessed his or her sins, he or she is absolved by the priest and receives the sacramental grace of Confession.[285]'

In a Roman Catholic church there will be a special Eucharist called a Requiem Mass where prayers are said for the dead person's soul.

HINDUISM

Like Buddhists, Hindus believe in reincarnation (samsara - the process of rebirth). The spirit simply moves from one body to another in its search for nirvana, something else they have in common. The bodies are also cremated because there is a belief that the burning releases the spirit from the body. Once again, the acts committed during life have an effect upon what happens after death. This is called Agami karma – 'the actions which are performed in the present life and which go on to affect the future'[286]. Paapa, the sinful actions of the past, can cause suffering. This is not caused by harming others, there is no unforgivable sin in Hinduism, but Paapa is the harm a person does to themselves though their actions.

Interestingly, suicide is not an entirely closed book in

---

[285] http://catholicism.about.com/od/thesacraments/g/Last_Rites.htm
[286] www.bbc.co.uk

Hinduism. They cannot permit themselves to die because their girlfriend has left them or because of depression. It must be because they believe that they are coming to the end of their physical life and, from this, they may withdraw diet and fluids. It is a renunciation of the world and of the flesh, which shows that the present state is temporary and unimportant.

As soon as death occurs, those gathered will avoid unnecessary touching of the body, as it is seen as impure.

> 'Preparations for the funeral begin immediately. The funeral should take place as soon as possible—traditionally, by the next dusk or dawn, whichever occurs first. A priest should be contacted and can help guide in the decision-making process and direct the family to a Hindu-friendly funeral home… Traditionally, all Hindus—except babies, children, and saints—are cremated…Traditionally, the ashes should be immersed in the Ganges River, though more and more other rivers are becoming acceptable substitutes. For Hindus living outside of India, there are companies that will arrange for the shipment of the cremated remains to India and will submerge the ashes in the Ganges.[287]'

This appeared in the Telegraph today, 08 October, 2016:

River Soar is new Ganges

By Karyn Miller

10 Oct 2004

'There are no temples or funeral pyres on its banks, but the River Soar in Leicestershire has been approved as an alternative holy site to India's

---

[287] https://www.everplans.com/articles/hindu-funeral-traditions

> Ganges for Hindu funerals.
> After requests from the Asian community of over 250,000, the Environment Agency has approved ceremonies in which the ashes of dead Hindus and Sikhs are scattered across the water.
> A boat hire company has also been authorised to provide a customised service for the funerals, which are increasing in demand.
> This week, Shastriji Prakashbhai Pandya, a Hindu priest who officiates the ceremonies, claimed that the Soar was an acceptable alternative. "When I close my eyes, this could be the Ganges," he said.
> The leap of imagination required to conjure up such a vision is large.'

I'll say.

JUDAISM

Judaism has been the target of hatred and abuse by other religions for millennia. This is based upon the fact that, unlike Christians, they can lend money, which is against Christian beliefs (neither a borrower nor a lender be etc).

> 'The Bible has a great deal to say about the power of money. In particular, it is quite specific about how we should treat debt and lending…at heart of the story of salvation we find the power of money and liberation from debt is a central concern. The admonition that we cannot serve both God and Mammon (Matt 6:19-24) it is not a trivial matter: the central drama of salvation history is an act of liberation from debt slavery. To put the pursuit of money before the welfare of people, and use money to re-enslave and exploit people, especially the poor and vulnerable, is to turn your back on God's salvation and deny in practice the revelation given

> in Scripture of who God is. Whereas to use money to serve the common good, and in particular to relieve the poor, is a mark of salvation.[288]'

If you get the chance to read or see *The Merchant of Venice*, it is worth it because it does display the very realistic attitudes of others towards Jews and reflects not only historic prejudices, but also the prejudice of the Elizabethan times, during which the play was written. I use prejudice, but perhaps bias might be more appropriate as the dislike of Jews is based upon a firm and 'educated' belief. It's a moot point. If you read (The Great) Bill Bryson's *1927*, you will see that antisemitism has survived quite thoroughly into the 20th and 21st centuries under the watchful eye of all those supposedly belief-tolerant, freedom-loving countries.

There is also the ever-present idea that it was the Jews that contributed heavily to the crucifixion of Jesus.

> 'According to the gospel accounts, Jewish authorities in Roman Judea charged Jesus with blasphemy and sought his execution (see Sanhedrin Trial of Jesus), but lacked the authority to have Jesus put to death (John 18:31), so they brought Jesus to Pontius Pilate, the Roman governor of the province…[289]'

The Jewish religion can be broken down into Orthodox, Reform or Conservative. However, all groups believe that when they die they will go to Heaven to be with God.

> 'This next world is called Olam HaEmet or 'the world of truth'. Death is seen as a part of life and a part of God's plan. The burial takes place as soon as possible following the death. Pallbearers will carry the casket to the grave. A family member will

---

[288] http://www.theology-centre.org.uk/neither-a-borrower-nor-a-lender-be/

[289] *Jewish Deicide.* https://en.wikipedia.org/wiki/Jewish_deicide

> throw a handful of earth in the casket with the body. This is to put the body in close contact with the earth. Jewish law says each grave must have a tombstone to remember the deceased.[290]'

ISLAM

There is very little difference between Islam and Christianity. There, I've said it. It's quite remarkable. For all the childish squabbles and intolerance that we perceive on the news every day, at their cores, these religions were born of the same mother.

Muslims believe that the soul continues after death and the fate of that soul can be shaped by what a person has done in their life. They also believe, as do Christians, that there will be a Day of Judgment by Allah, their name for God. Until the Day of Judgment, the deceased will remain in their grave until, on that day, Allah decides whether they will go to Heaven or to Hell.

A major difference is that the body must be buried within 24 hours of death as Muslims believe that the soul leaves the body at the moment of death.

You know, I just don't think it matters how you reach your God (in a nice way of course), as long as you get there.

SIKHISM

And again, we go into Buddhist/Hindu territory and the cycle of reincarnation. In Sikhism, a person's soul may be reborn many times as a human or an animal. Therefore, for Sikhs, death is not the end. The Sikh sacred text, the Guru Granth Sahib, says that the body is just clothing for the soul and is discarded at death.[291]

Disposal of the body is also by cremation.

---

[290] *An Outline of Different Cultural Beliefs at the Time of Death*. Loddon Mallee Regional Palliative Care Consortium.
[291] www.bbc.co.uk

ATHEISM

In these days, when we are told to respect everyone's beliefs (to the point of intolerance), we mustn't forget that some people choose not to believe and they choose not to believe not through laziness or because they just can't be arsed, but because they have, after much debate, research and thought, come to the belief that they do not want to have a belief.

I am working my way through *The God Delusion* by Richard Dawkins at the moment (as well as (The Great) Bill Bryson's *The Road to Little Dribbling* – how's that for a schizophrenic contrast?) and his rant makes me slightly scared. He seems very angry. I'm not saying that his anger isn't justified, indeed I share much of it, but, bloody hell, it is scathing.

I can understand why it is scathing. He is defending the right not to believe, the right to not be cowed by superstition, to not be manipulated by an ingrained, Machiavellian fear stoked by religion and our own fear of death; to be left the hell alone.

> 'The God of the Old Testament is arguably the most unpleasant character in all fiction: jealous and proud of it; a petty, unjust, unforgiving control-freak; a vindictive, bloodthirsty ethnic cleanser; a misogynistic, homophobic, racist, infanticidal, genocidal, filicidal, pestilential, megalomaniacal, sadomasochistic, capriciously malevolent bully... There is something infantile in the presumption that somebody else has a responsibility to give your life meaning and point...The truly adult view, by contrast, is that our life is as meaningful, as full and as wonderful as we choose to make it.'

I think his bite has more effect because we do now live in a prickly, contrary, politically correct society where free thought is condemned if it goes against the prevailing winds. We are as oppressive and intolerant and enslaving now as we have ever been in our desperate scrabble to set others free.

Anyone has the right to find their God in any way they choose, providing they do not show inconsideration or do harm to those with other beliefs or even no beliefs at all. That is tolerance.

PART 7

The Psychology Of Death

2016 has been the worst year that I can recall for the deaths of the well-known. To name but a few: Alan Rickman, David Bowie, Johann Cruyff, Prince, Glenn Frey, Cecil Parkinson, Frank Finlay, Terry Wogan, Boutros-Boutros Ghali, Umberto Eco, Harper Lee, Carla Lane, Douglas Slocombe, Lemmy, George Kennedy, Frank Kelly (feck it), Nancy Regan, Kenny Baker, Caroline Aherne, George Martin, Ken Adam, Frank Sinatra Jr, Paul Daniels, Gary Shandling, Ronnie Corbett, Guy Hamilton, Dave Swarbrick, Gene Wilder, Victoria Wood, Muhammad Ali and Anton Yelchin.

These are only a few, who were more significant to me. I'm not sure that *I'm* going to make it to the end of the year at this rate (it is, at this moment of amendment, December 1, 2018. Phew!)

I realise that this is a part of growing older. Those that you have 'grown up' with, so to speak, start to fall by the wayside. I thought Wogan and Wilder would be here forever. I will miss them.

But, it's not just about missing them. I don't just mourn them, because they have taken a part of me with them. I am forced to mourn my own gradual passing. I am forced to recognise my own mortality and, let's face it, the longer you live, the more likely you are to die.

I am forced, as many of us are when we read these headlines, to take account, however unwillingly, however subconsciously, of myself.

When Jon Lord, the keyboard player of Deep Purple died, I shed some tears. This was not just because he seemed like a remarkably nice and extraordinarily talented man, but because he and his fellow band members had been the soundtrack to a significant part of my life. His passing was, in part, my passing too.

This all sounds a little selfish and self-centred doesn't it, but it's not, I assure you. It is something we all go through at some time, whether we like to admit it or not. Some of us suppress it, most of us do, but once in a while we have to take ourselves off into a quiet corner and ruminate in that Very British Way.

It's not just death that we mourn either. I can remember that one of my nursing tutors (we had them in the good old days) liked to tell the story of her lost handbag as the preamble to her session on grief. Of course, we had all been warned by those in the years in front of us of the tale to come and, rather than being moved to tears of sadness by the tragedy of her missing bag, we were moved to tears of laughter.

She had a point though. When we lose something, whether it is your mother or your car keys, we always go through particular stages of grief, however minimal. Hopefully though, the loss of car keys doesn't end up with ten years of counselling and a sick note.

In 1969, Elizabeth Kübler-Ross, in *On Death and Dying*, wrote the new and, as it turns out, definitive, description of the stages of grief. It is difficult to overestimate the effect that her work had. She broke barriers and broke silence. Death has always been a taboo subject, particularly in the 'civilised' world, but she brought the subject into the open and made us not only examine our fears as a society, but also took the care of the dying to new levels.

She broke the subject down into five stages:

1. Denial
2. Anger
3. Bargaining
4. Depression
5. Acceptance

These are natural and acceptable reactions to dying. They might well be extreme and they might not progress neatly from one stage to another or even happen in this order, but my experience of working with dying people has found this to be helpful and pretty accurate.

Denial: When you are told bad news, the inclination, the instinct, is to deny it. We are all ostriches and most of us will want to dive headlong into the sand. This denial can lead to the affected individual becoming withdrawn, uncommunicative, short-tempered or they might even become elated. On the other hand, their emotions might become incredibly flat, almost non-existent.

Anger: It is quite natural that the individual will feel angry. They have been cheated. They have been told that the one thing that they thought they owned, is not actually theirs to keep. There is a need to blame, which is why sometimes doctors and other health professionals get sued or verbally/physically attacked.

I recall one patient who had cancer and the secondaries had wreaked havoc with his spine, to the point where he had become wheelchair bound. He was a cantankerous old bastard and his son was hacked out in his image, but I liked him because he said what he thought. My colleagues avoided him and when his son came to visit, they hid. I was volunteered to attend the multi-disciplinary team meetings because nobody else would.

Eventually we came to the point where this man had to move on, either to home or to some other place of care. There was nothing else we could do for him. His pain was controlled and the care he needed could be delivered elsewhere – we needed the bed.

He and his son never got past stage two. In the final multi-disciplinary team meeting we had, they were still in denial of the facts, despite the fact that we showed them the results of the scans and the spread of the disease. He went home and died a short time later. He left us in a state of anger. I didn't blame him. It was one of the most upsetting times I had as a nurse. It must have been a living nightmare for him.

Bargaining: 'Dear God, if you let me live, I will give one-third of my wages to the poor and I will volunteer to help the homeless at Christmas.' This is the bargain. This is when religion might well play a part. This is when belief might well collapse and send the individual all the way back to a previous stage or open up a new perception of life. But the person bargains with themselves too. It is the beginning, if they can get through it, of compromise, because the bargaining isn't just with God or whoever they believe in, it is a time of reflection and conversation with oneself and a step towards realisation of the truth.

Depression: And with that truth comes pain and the certain knowledge that there is no way back for the individual, that they are finite, that no one is lying to them and that the end is nigh. Feelings of isolation and helplessness abound, along with despondency and probably tears.

Acceptance:

> 'For the dying individual, this is the stage at which one attempts things for what they are, makes peace with the world, and makes the relationships with the deceased dignified. They accept the situation, and go on with their life, accepts the loss as a part of life, although it will be in a different way.'
>
> Elisabeth Kübler-Ross

It's a long journey to get to this place. Many don't make it and will not go gentle into that good night.

Much like Florence Nightingale, Elisabeth Kübler-Ross is being put through the ringer a bit at the moment in certain quarters. There is the claim that her stages do not allow for the individual, that there is no real evidence to say that these stages are actually real, that the stages are prescriptive rather than descriptive - that people should do this, rather than they might do this. Relatives and patients might expect these stages and when these stages are not met, then they're expectations are not met and the professionals become, in their eyes, incompetent liars. There is the thought that not everybody, due to environment, length of disease, social standing, wealth, might get a chance to experience these stages and that, once again, expectations are not met. As a professional, if you do not get to see these stages, you might become flummoxed and wonder where you are left standing with the patient. Your expectations again are not met because the stages are not met. This might be very difficult for those who are newly trained. Those with experience tend to know better.

However, those who criticise, in their turn, usually tend

to come up with a similar list to Kübler-Ross, so yahboo sucks to them.

What Elisabeth Kübler-Ross did was open a debate upon which further debate could be had. That in itself is one hell of an achievement and I never saw anything that could better it. It's still being used and taught by professionals now and is a good starting point for further research. We should always strive to improve upon what we find.

Those who are left in the unenviable position of having to watch someone die, also go through a grieving process very similar to that talked about by Elisabeth Kübler-Ross. It is given a couple more stages, though it is usually between five and seven stages and is almost identical in need and outcome at each stage.

These are generally called:

1. Shock and Denial
2. Pain and Guilt
3. Anger and Bargaining
4. Depression, Reflection, Loneliness
5. The Upward Turn
6. Reconstruction and Working Through
7. Acceptance and Hope[292]

You can see immediately the similarities between the two, the difference is that in this instance, the individual has to accept life not death and move on to continue living. This isn't always as easy as accepting a finite end or whatever the dying individual's belief system says, but once again, those coping mechanisms such as religion, those bargaining processes, those times of anger and isolation and plain fear exist as much for the mourner as for the mourned.

This is, of course, scratching the surface of the subject. There is how death affects our everyday life, how it makes people behave in war, how it is used as a weapon, as a bargaining tool and

---

[292] *Stages of Grief: Through the Process and Back to Life.* www.adams.edu/administration/hr/7-stages-of-grief.pdf - among many others.

as a method of state and personal control.

There is a whole lot more research to embrace out there, and it should be embraced but, as I have said, this is '*A Beginner's Guide…*', that is all.

INTERMISSION

'A quotation is a handy thing to have about, saving one the trouble of thinking for oneself, always a laborious business.'
A.A. Milne, *If I May*

'It's not that I'm afraid to die, I just don't want to be there when it happens.'

Woody Allen.

'I don't want to achieve immortality through my work, I want to achieve it through not dying.'

Woody Allen.

'There are worse things in life than death. Have you ever spent an evening with an insurance salesman?'

Woody Allen.

'It's funny how most people love the dead, once you're dead you're made for life.'

Jimi Hendrix.

'My grandmother was a very tough woman. She buried three husbands and two of them were just napping.'

Rita Rudner.

'The report of my death was an exaggeration.'

Mark Twain.

'A thing is not necessarily true because a man dies for it.'

Oscar Wilde

'I do not fear death. I had been dead for billions and billions of years before I was born, and had not suffered the slightest inconvenience from it.'

Mark Twain

'Life should not be a journey to the grave with the intention of arriving safely in a pretty and well preserved body, but rather to skid in broadside in a cloud of smoke, thoroughly used up, totally worn out, and loudly proclaiming 'Wow! What a Ride!'

Hunter S. Thompson, The Proud Highway: Saga of a Desperate Southern Gentleman

'You only live twice:
Once when you are born
And once when you look death in the face'

Ian Fleming, You Only Live Twice

'Sleep, those little slices of death — how I loathe them.'

Edgar Allan Poe

'Do not go gentle into that good night,
Old age should burn and rave at close of day;
Rage, rage against the dying of the light.'

Dylan Thomas, Do Not Go Gentle Into That Good Night

'Of all the ways to lose a person, death is the kindest.'

Ralph Waldo Emerson

'A recent survey stated that the average person's greatest fear is having to give a speech in public. Somehow this ranked even higher than death which was third on the list. So, you're telling me that at a funeral, most people would rather be the guy in the coffin than have to stand up and give a eulogy?'

Jerry Seinfeld

‘I want to die like my father, peacefully in his sleep, not screaming and terrified, like his passengers.’

Bob Monkhouse

'Heaven, as conventionally conceived, is a place so inane, so dull, so useless, so miserable, that nobody has ever ventured to describe a whole day in heaven, though plenty of people have described a day at the seaside!'

George Bernard Shaw

'I'm prepared to meet my maker. Whether my maker is prepared for the ordeal of meeting me is another matter.'

Winston Churchill

‘Life is pleasant.
Death is peaceful.
It's the transition that's troublesome’

Isaac Asimov

‘Life does not cease to be funny when people die any more than it ceases to be serious when people laugh.’

George Bernard Shaw

‘I told you I was sick.’

Tombstone of Spike Milligan

‘Died at the age of 102 at the hands of a justifiably outraged husband.’

From a tombstone.

'The Iraqis have hundreds of seasoned suicide bombers.'

Fox News report.

'Either this man is dead or my watch has stopped.'

Groucho Marx on checking a pulse in A Day at the Races.

'I took a test in Existentialism. I left all the answers blank and got 100.'

Woody Allen

'I was thrown out of college for cheating on the metaphysics exam; I looked into the soul of the boy sitting next to me.'

Woody Allen

'A single death is a tragedy, a million deaths is a statistic.'

Joseph Stalin

'Death will be a great relief. No more interviews.'

Katharine Hepburn

'If you die you're completely happy and your soul somewhere lives on. I'm not afraid of dying. Total peace after death, becoming someone else is the best hope I've got.'

Kurt Cobain

'Do not fear death so much, but rather the inadequate life.'

Bertolt Brecht

Death is the wish of some, the relief of many, and the end of all.

Seneca

Once the game is over, the king and the pawn go back into the same box.

Italian proverb

I don't believe in an after life, although I am bringing a change of underwear.

Woody Allen

And lastly, possibly the saddest quote of all:

'Dearest, I feel certain that I am going mad again. I feel we can't go through another of those terrible times. And I shan't recover this time. I begin to hear voices, and I can't concentrate. So I am doing what seems the best thing to do. You have given me the greatest possible happiness. You have been in every way all that anyone could be. I don't think two people could have been happier 'til this terrible disease came. I can't fight any longer. I know that I am spoiling your life, that without me you could work. And you will I know. You see I can't even write this properly. I can't read. What I want to say is I owe all the happiness of my life to you. You have been entirely patient with me and incredibly good. I want to say that – everybody knows it. If anybody could have saved me it would have been you. Everything has gone from me but the certainty of your goodness. I can't go on spoiling your life any longer. I don't think two people could have been happier than we have been. V.'

Virginia Woolf, suicide letter to her husband, Leonard. March 1941

END OF INTERMISSION

PART 8

Break On Through (To the Other Side)[293]

[293] The Doors

If you run out of things to talk about in the pub, just say: 'So, who's seen a ghost?' Then sit back and relax as your friends go bonkers.

Everyone has an opinion on this. I bet even some ghosts don't believe in themselves.

> 'MORE than two-thirds of Britons have had a brush with the paranormal - and over a third claim to have seen a GHOST, according to a new survey. The study, carried out to discover how many people believed in "paranormal activity" and would pay to see a clairvoyant, found a whopping 82 per cent of Britons believe in the supernatural. More than two thirds (68 per cent) said they had experienced some sort of supernatural event - and more than half (56 per cent) said they had been given a sign that a deceased loved one is present.'
>
> www.express.co.uk – July 27 2015

You have just said, to yourself, inside your head, echoing as if whispered in a cave, either, 'Wow, that's interesting' or 'What a bunch of fucking nutters!'

The afterlife is an incredibly divisive subject and yet, as we have seen, it is one of the foundation stones of some major religions in the sense of reincarnation. Profits in the church depend upon our unquestioning belief in a giant, all powerful ghost. Every day, many of us pray to this ghost or scream his name in frustration at the water cooler not working.

It would seem that we are going to take some convincing that ghosts do not exist. Why are we all so taken by this? Well of course, there is fear. We all have a fear or at least an uncertainty about death. There is the media. Ghosts make headlines. Death makes headlines. Then there is the proliferation of, mostly American, ghost programmes. We Brits however can claim the ground-breaking *Most Haunted* as a leader in its field. This really let the chicken out of the egg. Literature is littered with ghosts and the supernatural from Shelly to Shakespeare to Poe to King. We love it. We love it because we don't just love being scared, but because it makes us ask questions that deep down inside we are afraid to ask.

Visitations from the dead are all part of the process of grief, a natural part of grief. They can be triggered by smells, by sounds, by surroundings that were familiar to you as a couple or simply by need.

> 'Cleiren's Leiden study showed that fourteen months after a death about a third of the bereaved people studied felt a sense of the presence of the dead and also 'talked' to the dead either vocally or in a silent inner 'conversation".
>
> *Death, Ritual, and Belief: The Rhetoric of Funerary Rites* - Douglas Davies
>
> 'Mourning seems to be a time when hallucinations are particularly common, to the point where feeling the presence of the deceased is the norm rather than the exception. One study, by the researcher Agneta Grimby at the University of Goteborg, found that over 80 % of elderly people experience hallucinations associated with their dead partner one month after bereavement, as if their perception had yet to catch up with the knowledge of their beloved's passing. As a marker of how vivid such visions can seem, almost a third of the people reported that they spoke in response to their experiences. In other words, these weren't just peripheral illusions: they could evoke the very essence of the deceased.'
>
> www.scientificamerican.com

The fact that we know that these visions can be

stimulated by grief does not make them any less real. If you're interested in this, watch *Truly, Madly, Deeply* with the late Alan Rickman. It's a piece of fiction and yet a great study in the effects of grief.

There are also the visions that come at the time of death. Many of the nurses I have known, myself included, know of patients who are dying who have been visited by a relative.

Whether this is a trick of the mind, an unconscious way for the one dying to cope with their death by telling themselves that something better awaits them 'on the other side', is difficult to say. The brain is a swirl of chemicals, stimulated by both physical and psychological events and death is possibly the actual peak of both of those two sensations. There is no doubt though that great comfort has been found in these visions and that people have gone to their death in a more peaceful state of mind because of them.

I want to look, just quickly, at four areas which have a heavy influence on our society:

Reincarnation
Near Death Experiences (NDEs)
Ghosts
Mediums

I have no intention of trying to come to any conclusion as to whether or not they exist. There is only one way to do that. I'm just going to look at some of the evidence that is available for each and let the nibble either spark an appetite in you or make you vomit with the stupidity of it all.

Reincarnation

So, what exactly is reincarnation? It is the rebirth of the soul into another body. Some believe that this could be the body of a human or beast, some believe that it is just into the body of another human.

In religion, it is a part of that cycle which aims at achieving enlightenment. Once this has been achieved, the soul may rest.

Walter Semkiw, MD from www.iisis.net, says that there are nine Principals of Reincarnation:

> 'Independently researched cases demonstrate the following principles of reincarnation, which I will be citing as we go through our case studies.
>
> •Facial features are observed to be similar from one incarnation to another
> •Spirit being guidance is observed in the development of reincarnation cases
> •Relationships seem to be renewed through reincarnation. People are shown to reincarnate in soul or family groups and there is evidence that souls plan their incarnations, including who their parents and siblings will be.
> •Religion, race and ethnic affiliation can change from lifetime to lifetime, which has huge positive implications for society. Most conflicts are based on differences in these cultural markers of identity. To know that we can change religion, race and ethnic affiliation from one incarnation to another can help us see ourselves in more universal terms and thus create a more peaceful world.
> •Child prodigies and innate talent are explained through reincarnation.
> •Geographic memory is observed, which I define as past life memories or emotions that are stimulated by visiting past life locations
> •It is demonstrated that the soul can inhabit more than one body at a time, a phenomenon which I call "split incarnation." An older term is for split incarnation is parallel lives.
> •Xenoglossy cases demonstrate that past life personalities are retained intact within the soul. Xenoglossy refers to the phenomenon of an

individual being able to speak a foreign language that was not learned by normal means in the contemporary lifetime. In some xenoglossy cases, the past life personality is observed to emerge intact, sometimes not even knowing that it has died.
• Reincarnation cases are also observed where individuals are attracted to their own past lives, which I call "affinity cases.'"[294]

This is to a degree reinforced by www.reincarnation-research.com:

'In the past half a century, reincarnation has become a subject of scientific inquiry due to the availability of information that are verifiable. There are four main sources of evidence:

Authenticated memories of past lives
These are individuals who remember their past lives, and whose detailed memories have been independently verified to be correct.
Birthmarks
Those who exhibit body marks or characteristics that are connected with the previous life, such as scars, deformities or marks.
Xenoglossy
Those who remember and speak a language unlearned in the current life but which was known to the subject in a previous life.
Testimonies
Testimonies of famous people regarding their memories of a previous life.'

Semkiw bases a lot of his own work upon the work of Dr Ian Stevenson of the University of Virginia School of Medicine.

---

[294] http://www.iisis.net/index.php?page=walter-semkiw-ian-stevenson-society-for-scientific-exploration-sse&hl=en_US

Stevenson was Chair of the Department of Psychiatry from 1957 to 1967, the Carlson Professor of Psychiatry from 1967 to 2001, and a Research Professor of Psychiatry from 2002 until his death in 2007. His studies included reincarnation, near-death experiences, out-of-body experiences, after-death communications, deathbed visions, altered states of consciousness and psi (supposed parapsychological or psychic faculties or phenomena).

In his paper *Birthmarks and Birth Defects Corresponding to Wounds on Deceased Persons* (1992), he put forward the idea that birthmarks and birth defects were evidence of wounds carried forward from another life:

> 'About 35% of children who claim to remember previous lives have birthmarks and/or birth defects that they (or adult informants) attribute to wounds on a person whose life the child remembers.'

He is completely aware of the fragility of his case:

> 'Because most...of these cases develop among persons who believe in reincarnation, we should expect that the informants for the cases would interpret them as examples according with their belief; and they usually do. It is necessary...to think of alternative explanations. The most obvious explanation of these cases attributes the birthmark or birth defect on the child to chance, and the reports of the child's statements and unusual behavior then become a parental fiction intended to account for the birthmark (or birth defect) in terms of the culturally accepted belief in reincarnation. There are, however, important objections to this explanation. First, the parents (and other adults concerned in a case) have no need to invent and narrate details of a previous life in order to explain their child's lesion. Believing in

reincarnation, as most of them do, they are nearly always content to attribute the lesion to *some* event of a previous life without searching for a *particular* life with matching details. Second, the lives of the deceased persons figuring in the cases were of uneven quality both as to social status and commendable conduct. A few of them provided models of heroism or some other enviable quality but many of them lived in poverty or were otherwise unexemplary. Few parents would impose an identification with such persons on their children. Third, although in most cases the two families concerned were acquainted (or even related), I am confident that in at least 13 cases (among 210 carefully examined with regard to this matter) the two Families concerned had never even heard about each other before the case developed.'

and was rigorous in his methods:

'My investigations of these cases included interviews, often repeated, with the subject and with several or many other informants for both families. With rare exceptions, only first-hand informants were interviewed. All pertinent written records that existed, particularly death certificates and post-mortem reports, were sought and examined. In the cases in which the informants said that the two families had no previous acquaintance, I made every effort to exclude all possibility that some information might nevertheless have passed normally to the child, perhaps through a half-forgotten mutual acquaintance of the two families. 1 have published elsewhere full details about methods (Stevenson, 1975; 1987). I did not accept any indicated mark as a birthmark unless a first-hand witness assured me that it had been noticed

> immediately after the child's birth or, at most, within a few weeks. I enquired about the occurrence of similar birthmarks in other members of the family; in nearly every instance this was denied, but in seven cases a genetic factor could not be excluded. Birth defects of the kind in question here would be noticed immediately after the child's birth. Inquiries in these cases excluded (again with rare exceptions) the known causes of birth defects, such as close biological relationship of the parents (consanguinity), viral infections in the subject's mother during her pregnancy, and chemical causes of birth defects like alcohol.'

It is the thoroughness and the reserve in approach that impresses, the willingness to accept other causes and to balance those other causes with the *possibility* of reincarnation.

There is a particular interest in the role of children in the concept of reincarnation. Stevenson prefers to use children because their memories of their previous experiences are usually fresher and they have not been compromised by the cynicism and stories that come with age. There is also a clarity that comes with youth which fades and, as the person grows older and accepts their present life, those memories might well fade altogether.

It is difficult to forget the disaster that was regression therapy, so popular during the seventies and eighties which was eventually tainted by the rather obvious idea that people were prejudiced by events and concepts that they had come across in their present lives. It was intended to find a cause for deep-rooted problems within this life within a past life. The pitfall presented is so blatant that I wonder how people cannot have noticed this in the first place, although the shine of a coin has been known to blind.

Jim B Tucker, an associate psychiatry professor at the University of Virginia Medical Center's Division of Perpetual Studies (there's a name for you. Also associated with Stevenson's work) has been studying the subject for more than fifteen years on children who are usually between the ages of two and six. 'The children are sometimes able to provide enough detail about those lives that their stories can be traced back to an actual person—rarely famous and often entirely unknown to the family—who died years before[295]'.

Tucker claims that 70% of the children say that they died a violent or unexpected death in their previous life and that males account for three-quarters of those deaths and that 'nearly 20% of the children studied have scarlike birthmarks or even unusual deformities that closely match marks or injuries the person whose life the child recalls received at or near his or her death.' 20% claimed to remember the time between death and rebirth, the 'intermission', but there is little consistency in the reports.

Tucker's Stats:

- The median age at the time of reincarnation is 28.
- 60% of children who claim past-life memories are male.
- in those cases, just over 70% of the deceased individuals claim they died a violent or unnatural death.
- 90% of the children say they were the same sex in a previous life as they are now
- The median time between claimed death and birth is 16 months.
- 20% of children claim memories of the time between death and rebirth.
- More children claiming memories of a past life between the ages of 2 and 6.

Tucker does not dismiss the alternatives either, such as fantasy play, fraud and faulty memories.

If you wish to know a little more about Dr Stevenson,

---

295 *The Science of Reincarnation.* Sean Lyons. 2013.

there is a fabulous *Omni* interview with him at http://reluctant-messenger.com/reincarnation-proof.htm.

I'm putting these ideas forward, not as a convert, but as one of open mind. I am one of those annoying individuals that will not believe something without evidence, no matter how much (in the words of Mulder) I want to believe. If there is the possibility of any of these supernatural occurrences being real, then it makes it worth investigating, if only because we do not yet entirely grasp quantum physics or the capacity of the brain to create, real or unreal, altered states.

I would like to finish this section with the story of a lady named Shanti Devi, born in 1926, in India. If you refer back to Semkiw's Principals of Reincarnation, then you will see common ground. You do have to bear in mind the chicken and the egg, however, and the fact that the two might have appeared spontaneously at the same time.

- 'On January 18, 1902, Chaturbhuj, a resident of Mathura, had a daughter, who was named Lugdi.
- When Lugdi reached the age of 10, she was married to Kedarnath Chaube, a shopkeeper of the same locality. It was the second marriage for Kedarnath. His previous wife had died.
- Lugdi was very religious. While on one pilgrimage, she was injured in her leg for which she had to be treated, both at Mathura and later at Agra.
- When Lugdi became pregnant for the first time, her child was stillborn following a Caesarean section. For her second pregnancy, her husband took her to the government hospital at Agra, where a son was born, again through a Caesarean on September 25, 1925. Nine days later, on

October 4, Lugdi's condition deteriorated and she died.

- One year ten months and seven days after Lugdi's death, on December 11, 1926, Babu Rang Bahadur Mathur of Chirawala Mohulla, Delhi, had a daughter, named Shanti Devi.
- Until the age of four she did not speak much. But when she started talking, she talked about her 'husband' and her 'children.'
- She said that her husband was in Mathura where he owned a cloth shop and they had a son. She called herself Chaubine (Chaube's wife).
- She narrated a number of incidents connected with her life in Mathura with her husband. On occasions at meals, she would say, 'In my house in Mathura, I ate different kinds of sweets.' Sometimes she would tell what type of dresses she used to wear. She mentioned three distinctive features about her husband: he was fair, had a big wart on his left cheek, and wore reading glasses. She also mentioned that her husband's shop was located in front of Dwarkadhish temple.
- By the age of six Shanti Devi even had a detailed account of her death following childbirth. Her parents, concerned at her persistence in the story, consulted their family physician, who was amazed how a little girl narrated so many details of the complicated surgical procedures (for the leg injured on pilgrimage).
- As the girl grew older, she kept asking her

parents to take her to Mathura. She, however, never mentioned her husband's name up to the age of eight or nine. It is customary in India that wives do not utter the name of their husbands. She would say that she would recognise him, if taken there.

- One day a distant relation, Babu Bishanchand, told Shanti Devi that if she told him her husband's name, he would take her to Mathura. She told him the name Pandit Kedarnath Chaube. Bishanchand then told her that he would arrange for the trip to Mathura. He wrote a letter to Pandit Kedarnath Chaube (her husband from the previous life), detailing all the statements made by Shanti Devi, and asked him to visit Delhi. Kedarnath replied confirming most of her statements and suggested that one of his relatives, Pandit Kanjimal, who lived in Delhi, be allowed to meet this girl.
- At the meeting, Shanti Devi recognized him as her husband's cousin.
- She gave some details about her house in Mathura and informed him of the location where she had buried some money. When asked whether she was able go by herself from the railway station to her house in Mathura, she replied in the affirmative, if they would take her there.
- Kanjimal was so impressed that he went to Mathura to persuade Kedarnath to visit Delhi. Kedarnath came to Delhi on November 12, 1935, with Lugdi's (Shanti Devi's) son Navneet Lal and his present

wife.

- They went to Rang Bahadur's (Shanti Ravi's parental home) house the next day.
- To mislead Shanti Devi, Kanjimal introduced Kedarnath as the latter's elder brother. Shanti Devi blushed and stood on one side. Someone asked why she was blushing in front of her husband's elder brother. Shanti said in a low firm voice, 'No, he is not my husband's brother. He is my husband himself.' Then she addressed her mother, 'Didn't I tell you that he is fair and he has a wart on the left side cheek near his ear?'
- When the mother asked what she should prepare as a meal for the guests, Shanti Devi said that her 'husband' was fond of stuffed potato parathas and pumpkin squash. Kedarnath was dumbfounded as these were his favourite dishes.
- Kedarnath asked whether she could tell them anything unusual to establish full faith in her. Shanti replied, 'Yes, there is a well in the courtyard of our house, where I used to take my bath.'
- She recognised her son as her son immediately. When asked how she was able to recognise him (she had died not long after giving birth to him), her reply, enigmatically, was that 'her son was a part of her soul and the soul is able to easily recognize this fact.'
- After dinner, Shanti asked Kedarnath, 'Why did you marry her?' referring to his present wife. 'Had we not decided that you will not remarry?' Kedarnath had no reply.

- Shanti Devi went to Mathura (the birth and death place of Lugdi) on November 24, 1935.
- At the station, an older man, wearing a typical Mathura dress, whom she had never met before, appeared front of her, mixed in the small crowd, and paused for a while. She was asked whether she could recognise him. His presence reacted so quickly on her that she at once touched the stranger's feet with deep veneration and stood aside. On inquiring, she whispered that the person was her 'Jeth' (older brother of her husband - aka brother-in-law). The man was Babu Ram Chaubey, who was really the elder brother of Kedarnath Chaubey.
- On the way to her house, she described the changes that had taken place since her time, which were all correct. She recognized some of the important landmarks which she had mentioned earlier without having been there. She guided the driver to her house without any difficulty.
- As they neared the house, she…noticed an elderly person in the crowd. She immediately bowed to him and told others that he was her father-in-law, which was true. When she reached the front of her house, she went in without any hesitation and was able to locate her bedroom. She also recognized many items of hers. She was tested by being asked where the 'jajroo' (lavatory) was, and she told where it was. She was asked what was meant by

'katora.' She correctly said that it meant paratha (a type of fried pancake). Both words are prevalent only in the Chaubes of Mathura and no outsider would normally know of them.

- She was asked her about the well of which she had talked in Delhi. She ran in one direction; but, not finding a well there, she was confused. Even then she said with some conviction that there was a well there. Kedarnath removed a stone at that spot and, sure enough, they found a well. As for the buried money, Shanti Devi took the party to the second floor and showed them a spot where they found a flower pot but no money. The girl, however, insisted that the money was there. Kedarnath later confessed that he had taken out the money after Lugdi's death.
- She was taken to her parents' home, where at first she identified her aunt as her mother, but soon corrected her mistake. She also recognized her father.
- Shanti Devi was then taken to Dwarkadhish temple and to other places she had talked of earlier and almost all her statements were verified.
- Lugdi's brother told me that Shanti Devi remembered her old friends and inquired about them. Similarly, Lugdi's sister informed me that Shanti Devi told a number of womenfolk about Lugdi having lent them some money, which they accepted as true. Shanti's emotional reactions on meeting relatives from her previous life were very significant.

- A friend of Kedarnath, 72-year-old Pandit Ramnath Chaube, told of a very significant event. When Kedarnath was in Delhi to meet Shanti Devi, he stayed at Pandit Ramnath Chaube's place for one night. Everyone had gone to retire, and only Kedarnath, his wife, his son Navneet, and Shanti were in the room; Navneet was fast asleep. Kedarnath asked Shanti that when she was suffering from arthritis and could not get up, how she became pregnant. She described the whole process of intercourse with him, which left Kedarnath in no doubt that Shanti was his wife Lugdi in her previous life. When this incident [was mentioned] to Shanti Devi, she said, 'Yes, that is what fully convinced him.'
- Dr Ian Stevenson 'also interviewed Shanti Devi, her father, and other pertinent witnesses, including Kedarnath, the husband claimed in her previous life. My research indicates that she made at least 24 statements of her memories that matched the verified facts.'[296]

I have presented what I see as only the pertinent facts and have cross-checked them with the sources mentioned below. The danger of course is that the sources could all spring from the same well.

---

[296] Sources:
http://ocoy.org/the-thrilling-narrative-of-shanti-devi-and-her-past-life/
www.beliefnet.com/faiths/hinduism/2002/06/i-have-lived-before-the-reincarnation-of-shanti-devi.aspx
The Case of Shanti Devi by Dr. K.S. Rawat/ www.carolbowman.com
[296] http://explaininglifesmysteries.blogspot.co.uk/2009/06/former-life-of-shanti-devi.html

I find it to be an extraordinary tale, but cannot verify the truth of it, only the consistent retelling and the evidence that comes with it.[296]

Near Death Experiences (NDEs)

According to the International Association for Near Death Studies (IANDS), there are certain common features to all NDEs:

- 'Intense emotions: commonly of profound peace, well-being, love; others marked by fear, horror, loss
- A perception of seeing one's body from above (called an out-of-body experience or OBE), sometimes watching medical resuscitation efforts or moving instantaneously to other places
- Rapid movement through darkness, often toward an indescribable light
- A sense of being "somewhere else," in a landscape that may seem like a spiritual realm or world
- Incredibly rapid, sharp thinking and observations
- Encounter with deceased loved ones, possibly sacred figures (the Judges, Jesus, a saint) or unrecognized beings, with whom communication is mind-to-mind; these figures may seem consoling, loving, or terrifying
- A life review, reliving actions and feeling their emotional impact on others
- In some cases, a flood of knowledge about life and the nature of the universe

- Sometimes a decision to return to the body
- A sensation of floating out of one's body. Often followed by an out-of-body experience where all that goes on around the "vacated" body is both seen and heard accurately.
- Passing through a dark tunnel. Or black hole or encountering some kind of darkness. This is often accompanied by a feeling or sensation of movement or acceleration. "Wind" may be heard or felt.
- Ascending toward a light at the end of the darkness. A light of incredible brilliance, with the possibility of seeing people, animals, plants, lush outdoors, and even cities within the light.
- Greeted by friendly voices, people or beings who may be strangers, loved ones, or religious figures. Conversation can ensue, information or a message may be given.
- Seeing a panoramic review of the life just lived, from birth to death or in reverse order, sometimes becoming a reliving of the life rather than a dispassionate viewing. The person's life can be reviewed in its entirety or in segments. This is usually accompanied by a feeling or need to assess loss or gains during the life to determine what was learned or not learned. Other beings can take part in this judgment like process or offer advice.
- A reluctance to return to the earth plane, but invariably realizing either their job on earth is not finished or a mission must yet be accomplished before they can return to

stay.

- Warped sense of time and space. Discovering time and space do not exist, losing the need to recognize measurements of life either as valid or necessary.
- Disappointment at being revived. Often feeling a need to shrink or somehow squeeze to fit back in to the physical body. There can be unpleasantness, even anger or tears at the realization they are now back in their bodies and no longer on "The Other Side."

> Whether the NDE was beautiful or terrifying, near-death experiencers commonly say it was unlike a dream, "more real than real," the most powerful event in their lives. They struggle to find words to describe it, but insist they now know something new about reality, that "there's more than what's here" (in the physical world). Most feel deeply changed in their attitudes toward life, work, and relationships.[297]'

Certainly, many descriptions have involved standing in or at the entrance to a garden and the appearance of a significant other, a mother or mother-in-law or dead partner. Likewise, there have been some horrific accounts where people have viewed Hell itself or have had the sensation of drowning or being threatened by an unknown figure. Many related the fact that they are told that 'it's not your time' and have to return to their body.

Whichever of these is their reality, they speak of it as a completely different experience to a dream, they just *know* the difference, and describe it as 'real'. People commonly talk about being transformed by the experience, going on to change their jobs

---

[297] *Characteristics of a Near-Death Experience.* http://iands.org/ndes/about-ndes/characteristics.html

and lifestyles and often entering a profession where they help people. That does sound a little sickly sweet, I know, but these are the perceptions of those to whom the event has occurred.

> 'Roughly 5 percent of the general population and 10 percent of cardiac-arrest victims report near-death experiences…Across cultures and religions, people describe similar themes: being out of body; passing through a tunnel, river or door toward warm, glowing light; seeing dead loved ones greet them; and being called back to their bodies or told it's not time to go yet…"People are transformed forever by the experience," said Vanessa Charland-Verville, a neuropsychologist in the Coma Science Group at the University of Liege in Belgium. "People say they're more empathic, they changed jobs, they're giving, they want to help the planet."'[298]

There are really a couple of major studies that have hit the headlines with regard to NDEs. One is a five year study by Dr Penny Sartori, who at the time was an intensive care nurse. The other was by Southampton University, led by Sam Parnia and part of the AWARE study - *AWAreness during Resuscitation - A Prospective Study*, which took place in 15 hospitals in the United Kingdom, United States and Austria.

The AWARE study I have found to be inconclusive and disappointing, from a NDE perspective. They looked at 2060 patients in Austria, the UK and the US. Only 140 survivors completed the first stage of interviews. 101 of those 140 completed the stage two interviews.

> '46% had memories with 7 major cognitive themes: fear; animals/plants; bright light;

---

[298] www.livescience.com/28472-near-death-experiences-vivid.html. *Near-Death Experiences More Vivid Than Real Life*. Tia Ghose. April 5, 2013.

> violence/persecution; deja-vu; family; recalling events post-CA [cardiac arrest] and 9% had NDEs, while 2% described awareness with explicit recall of 'seeing' and 'hearing' actual events related to their resuscitation. One had a verifiable period of conscious awareness during which time cerebral function was not expected…
> Conclusions
> CA survivors commonly experience a broad range of cognitive themes, with 2% exhibiting full awareness. This supports other recent studies that have indicated consciousness may be present despite clinically undetectable consciousness. This together with fearful experiences may contribute to PTSD and other cognitive deficits post CA.'[299]

The conclusion is startlingly inconclusive. We know that people in a state of coma or unconsciousness retain their hearing. We are, whether we like it or not, influenced by all that we have seen and heard during our lives. The themes that run through a cardiac arrest - the heart massage, the defib, the delivery of medication - have been shown a hundred thousand times on fictional and reality TV, in movies, described in books and magazines. We can expect some of these memories to surface and then contribute to what our expectations of the event might be. If we think that we might view ourselves from above, well, we know what we look like, events are backed up by sound (for all we know hearing might become incredibly acute during the event), it doesn't really take a leap of the imagination to put the vision of expectation together with the sounds of reality.

The work done by Penny Sartori is fascinating, especially if you hear her talk about it with the conviction that she does. She says she is a scientist and is one of those (like me) who needs proof before proceeding on to belief. Yet she says that the study changed her perceptions on life and that she is convinced that there is

---

[299] www.sciencedirect.com/science/article/pii/S0300957214007394

something beyond death.

> 'I began my eight-year study as a cynic. But by the time it ended, I was convinced that near-death experiences are a genuine phenomenon.
> So, what exactly is a near-death experience? At its simplest, it's a clear and memorable vision that occurs when people are close to death — though only a small percentage of us will have one.
> Researchers now agree that each vision will contain at least one of several recognised components, such as travelling down a tunnel towards a bright light, meeting dead relatives, or having an out-of-body experience.
> As the person 'leaves' his body, he may hear a buzzing, whistling, whirring or humming sound, or a click. Another common component of NDEs is a beautiful garden with lush green grass and vividly coloured flowers. There may be a stream or river in the background. Some people enter the garden, while others reach a gate or barrier — and know that they'll die if they go through it.
> Throughout an NDE, hearing and sight become more acute, and awareness is heightened. Often, the experience has been described to me as 'realer than real'.
> As oxygen levels reduce in the blood, the brain becomes increasingly disorientated, confused and disorganised.
> Time ceases to have meaning. In many cases, it feels as if the vision has lasted for hours though the person may have been unconscious for only a few seconds or minutes. Sometimes, it feels as if time speeds up; sometimes it goes slower.'[300]

[300] www.dailymail.co.uk. 25 January 2014

She's not daft, not by a long way, and understands the potential interference of drugs upon the system, what the lack of oxygen can do and that the body produces chemicals, some of which are opioids, of its own during such times. But, she says that in several of the cases, oxygen levels in the blood were taken during the event and proved to be normal and that in some cases, no drugs were given.

In *There Is Nothing Paranormal About Near-Death Experiences: How Neuroscience Can Explain Seeing Bright Lights, Meeting The Dead, Or Being Convinced You Are One Of Them* by Dean Mobbs and Caroline Watt, they set out to 'debunk' some of the experiences people have during NDEs.

> 'Approximately 3% of Americans declare to have had a near-death experience. These experiences classically involve the feeling that one's soul has left the body, approaches a bright light and goes to another reality, where love and bliss are all encompassing. Contrary to popular belief, research suggests that there is nothing paranormal about these experiences. Instead, near-death experiences are the manifestation of normal brain function gone awry, during a traumatic, and sometimes harmless, event.'

Gloves off – ding, ding.

Many people report an awareness of being dead during the event – '50% reported an awareness of being dead, 24% said that they had had an out-of-body experience, 31% remembered moving through a tunnel, and 32% reported meeting with deceased people…it is a common anecdote that near-death experiences are associated with feelings of euphoria and bliss, only 56% associated the experience with such positive emotions, and some even reported negative experiences.' Mobbs and Watt compared the feeling to Cotard Syndrome, also known as 'Walking Corpse' Syndrome, where the individual has a feeling that they are dead. It has been associated with certain areas of the brain affected by trauma. It is unknown, they say, as to why Cotard Syndrome occurs, but suggest that it 'may simply be an attempt to make sense of the strange experiences that the patient is having.'

The paper says that the out-of-body experience (OBE), according to Canadian neurosurgeon Wilder Penfield, is brain-based. The same experiences can happen during interrupted sleep or with hypnogogia, 'vivid dreamlike auditory, visual, or tactile hallucinations associated with the wake-sleep cycle', which can result in the feeling that one has floated out of one's body.

OBEs can also be artificially induced by stimulating the brain[301]. The subjects of the experiments reported sensations similar to OBEs.

The tunnel of light, the favourite of film makers and drama queens alike, might well be down to poor perfusion to the eyes as blood is withdrawn from the peripheral system.

> 'Indeed, such tunnel vision is associated with extreme fear and hypoxia (i.e. oxygen loss), two processes common to dying.'

Meeting those who have already died is once again put down to the brain.

> 'Many neuroscientific studies have shown that brain pathology can lead to similar visions. For example, patients with Alzheimer's or progressive Parkinson's disease can have vivid hallucinations of ghosts or even monsters. Patients have also been noted as seeing headless corpses and dead relatives in the house, which has been linked to…lesions, suggesting that this result from abnormal dopamine functioning, a neurochemical that can evoke hallucinations.'

Electrical stimulation can produce a sensation of

---

301 Blanke, O. et al. (2002). *Stimulating Illusory Own-Body Perceptions.*

presence. Macular degeneration can produce hallucinations as can the brains of older people as certain areas of the brain try to compensate for nearby damaged areas.

The positive emotions, even elation, or the fears that sometimes come with NDEs, can be put down to the massive chemical upheaval within the body. The truth is, no one knows.

There has to be room here for the impact of expectations. We get what we expect. We expect to see Grandma, so we do. We expect the beauty of nature, so we get it. We expect to be found out for our sins and so we are threatened and punished accordingly.

There are common factors, that is undeniable, but the one major common factor in all of this is the human body and the template from which it comes. Do we have the same experiences because we are made the same, have the same physiology and psychology (yes we do, essentially) and therefore have the same results from our shared experiences? Do we understand so little of the brain that we have to put the experience down to the 'supernatural' because that is all we have? In a hundred years, the supernatural might well no longer exist, because we will have all the answers.

That's how God died, after all.

And what about the spirit? No one has concluded beyond a shadow of a doubt that we have a spirit. There is no doubt that the *spiritual* as a concept exists, a way for us to elevate ourselves above the impurities of this physical body, but whether something ethereal is able to leave our corporal prisons and wander off to see Grandma or be reborn in the body of a rat is all but unprovable.

However, that won't stop me reading about it and watching television programmes about it. I want there to be more, I really do. Am I no more than an ant? Are you? Are we really just vessels for genes to be passed on over the centuries to ensure the survival of that gene? Or are the gene and the soul separate entities that collide generation after generation in an attempt to reach their own form of enlightenment?

What I cannot, in my experience, deny, is that I have seen patients reaching out to someone who isn't there, saying that a long-dead relative is in the room.

A Japanese survey found that

> 'Of 2,221 survey responses, the researchers found that *Omukae* (literally, someone visiting a dying patient to accompany them on death's journey) were reported in 463 cases (21%)., 351 of the families stated that the patients themselves clearly described the deathbed vision, while 113 noted that, while the patient did not mention the vision, family members themselves witnessed the patient experiencing the phenomenon. Of the non-deathbed vision responses, 1,392 families reported no experience as occurring, while 365 families replied that they were unsure. Of the patients who did experience deathbed visions, 87% had visions of deceased persons (most often parents), while 54% had visions of afterlife scenes. Some of the interesting findings were that deathbed visions were significantly more likely to be observed in older patients and female patients, and in families with more religious activities, or who believed the soul survive the body after death.'[302]

There is a story my wife tells of one of her patients.

He lived in a nursing home. He was chronically ill and was on borrowed time. His daughter came from America to visit him, aware that this was probably the last time she would see him. When the visit was over, she returned home. On going to bed that night, she said to her deceased mother that it was okay, that she could now take him.

In the nursing home, the man got up, fell and broke his hip. When the staff asked him why he had got up at all, he had said that his wife had come to get him and that he was going with

---

302 www.dailygrail.com/Spirit-World/2016/9/Japanese-Study-Finds-1-5-Dying-Patients-Experienced-Deathbed-Visions-Deceased

her. Although ill, the man was *compos mentis* and normally mobile. He died shortly after.

Whether this is coincidence or not is impossible to say, but the timing of the events was so close together that, in the realm of NDEs, it has to qualify as evidence. It would certainly have qualified as evidence within the boundaries of the Japanese survey.

Ghosts

In all the years that I was a nurse, in all the hospitals in which I have worked, with all the people I have seen on their journey out of life, in all the years that I have lived, I have never seen a ghost.

You would think that the odds were in my favour.

I mean, I have been in the room with someone who has taken their final breath, I have washed and wrapped the bodies, I have taken bodies to the mortuary and yet not a thing, not a spooky whisper.

My eldest daughter has seen ghosts. When she was little, she was quite sensitive. She was with her mother one day at Cawthorne Park in Barnsley and she saw the ghost of a little girl upon the grass, near the big house, and waved at her. My wife asked to whom she was waving and she answered that she was waving to the little girl. Needless to say, there was no little girl there. Another time, they were in Meadowhall shopping centre (or The Five-Minute-Rent-Centre as it is now known) and her balloon suddenly burst. She asked her mum why the man had burst her balloon. There was no man anywhere near them - they were in the loo.

My wife, probably the most down to earth person in the whole wide world, recounted these tales and I believe her.

Despite my interest in the subject, I follow *Most Haunted*, *The Dead Files* and *Ghost Adventures* and read as much as I can find about the subject, I have never been presented with concrete evidence. The obvious and very justified argument is that these shows are entertainment and at no time do they present themselves as anything other. There is even an advisory comes up before the shows warning me (me personally) that the show is for entertainment purposes only.

I'm not actually sure what concrete evidence is. I think that it would have to be first hand, without the risk of camera trickery or Photoshoppery (that is a real word) and I would have to be sober, have the proper nutrition inside me and not be in a condition that might subject me to hallucinations, such as excessive tiredness or being in the middle of an enormous puff-high.

Is it possible to present evidence, even to the degree that evidence is presented for NDEs or OBEs?

On the basis that we are relying completely upon the truth of witnesses, then yes, I suppose it is. None of the NDEers came back from their journey with anything other than their word. The only thing I can say for sure is that it is no longer enough to simply rely upon photographic evidence or filmed evidence. There is too much scope for fakery.

What is for certain is that a whole industry has built up around our longing for the afterlife.

This fascination with the subject is as old as humankind. The Egyptians undoubtedly believed in an afterlife – evidence of this abounds. The same could be said of ancient Greece and Rome where gods were plentiful, along with the idea of placating them in order to ensure a smooth passage to the other side.

There was always a fear and a sense, an innate sense, that there was a parallel world to this one in some form or other. Shakespeare played upon it wholeheartedly as a form of doom-mongering and portent in his plays and Arthur Conan Doyle, the creator of the most logical of literary heroes, Sherlock Holmes, was an enthusiastic advocate of the paranormal.

Thirty-one per cent of Brits think they've seen, or felt, a ghost

Alison Lynch. Metro. Thursday 23 Jul 2015.

More than HALF of Britons claim to have had

contact with GHOSTS
MORE than two-thirds of Britons have had a brush with the paranormal - and over a third claim to have seen a GHOST, according to a new survey.

Jon Austin. www.express.co.uk. Mon, Jul 27, 2015

34% of British people believe in ghosts – and 9% have communicated with the dead

www.yougov.co.uk. 31 October 2014

America's most haunted: A fifth of people claim to have seen a ghost while 29% 'have made contact with the dead'
Women more likely than men to have seen apparitions and been in touch with the afterlife
Paranormal expert said many apparent hauntings depend on the individual's psychological state

Daily Mail. 31 October 2013.

That industry, sensing that the fox was in the trap and ready for the killing, pounced upon our needs like a pack of hounds, but instead of tearing our throats out, it gave us technology and the ability, helped by our own gullibility, to throw ourselves headlong into a fevered and desperate examination of the paranormal.

With the rapid advances in first garage and then commercial technology, we are induced to keep up the emptying of our wallets in an attempt to get closer to the other side.

These were the things that were offered, on one website, to help you find your very own Casper:

GADGET	PRICE
SB7 Spirit Box	£69.99

IR Thermometer	£8.15
Motion Detector	£8.41
EMF Detector	£44.99
Camcorder (with Night Vision)	£69.99
Notepad & Pens	£3
Walkie Talkies	£18.50
Extra Batteries (Pack of two)	£3.80
Torches (Ghost Hunting UV Torch)	£7.50
Digital Recorder	£13.65
Camera	£299.00
Camera Battery Pack	£7.25
TOTAL	**£554.23**

On top of this were offered ghost hunts, haunted accommodation and sleepover events.

Now, if you're anything like me, you might well have swapped your happy hat for that dark, cynical judge's wig. There is so much money involved in this business now that it is worth people's time to perpetuate the myth. The Paranormal Activity movies (the original cost $15000 to make) alone have grossed $401,363,355[303]. There's not a tremendous amount of information available as to how much the paranormal technology industry is worth, but I'm willing to bet it's on the up and up.

But what about the facts? Are there any?

Well, in this day and age, facts are hard to come by. Why? Because technology, despite all the ways in which it helps us in our everyday lives and in our health, wealth and fullness of existence, is a perfect tool for fraud. Even in Conan Doyle's time, the propensity for fraud was enormous. Though camera technology was in its youth, this didn't stop its exploitation.

303 www.boxofficemojo.com

The picture to the left was created by 'photographers Craig and George Falconers, a pair of charlatans who worked in the early part of the 20th century'. It is patently fraudulent, but more so judged from our 21st century techno-savvy perspective. From a 19th century perspective, it must have blown their socks off. This was not possible with the trickery of the camera! There

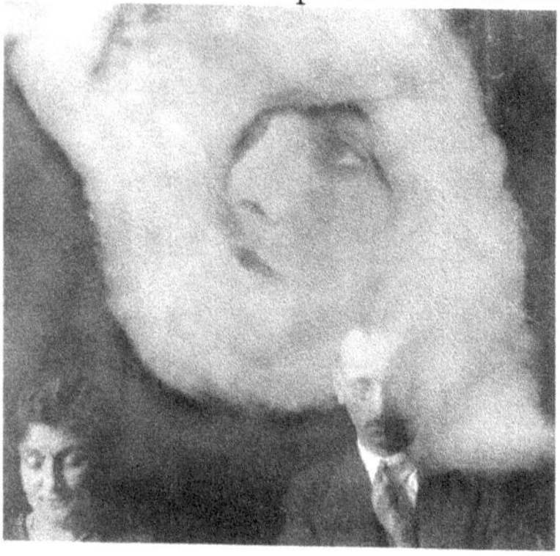

was a lack of cynicism and a sense of wonder at technological evolution. Today we accept Photoshop's abilities – it is only limited by the abilities and the imagination of the user – and there is now an assumption that anything that even hints at the paranormal was software-created. Perhaps that is actually a healthy, quite wise cynicism.

www.assap.ac.uk comes up with a very good point. Most ghosthunting is assumption led. There is the assumption when looking at this area to assume that there is an area to look at in the first place, so with the introduction of newer and better technology, there is the assumption that they have been introduced because they are there to find something that actually exists.

The assumptions that the site lists include:

Technique	Assumption
Use of mediums	assumes that mediums can contact ghosts
Calling out	assumes ghosts can hear and react
Baselines at start of vigil	assumes instrumental readings at the start of a vigil are 'normal'

Using Ouija, séances photos	assumes orbs are paranormal and associated with ghosts assumes ghosts can be contacted by these methods
Taking orb photos	assumes orbs are paranormal and associated with ghosts
EMF meters to detect ghosts	assumes ghosts can be detected by EMF meters
Dowsing for ghosts	assumes ghosts can be dowsed
Researching former inhabitants	assumes ghosts are former inhabitants of haunted site
Vigils in graveyards	assumes ghosts are more common in graveyards than elsewhere
Trying to record EVP	assumes ghosts can be contacted with EVP
Using instruments to ask questions	assumes ghosts can manipulate devices like torches, EMF meters, etc
Holding vigils in the dark	assumes ghosts are easier to detect in the dark

Because we are taking on these assumptions, does this influence our interpretation of any findings? Of course. You are more likely to hear something in the dead of night because the rest of the world has gone to sleep, therefore you can hear those creaks and groans that you don't hear during the day. Because of the particular environment that you are in, i.e. 'a haunted house', and your psychological state at the time, you assume that the noise is paranormal.

The spirit boxes are another great example of wishful/accidental/genuine/fraudulent misinterpretation.

Spirit boxes scan the airwaves for white noise in the assumption that spirits are able to contact us through white noise. That in itself is a big assumption. When the interviewer says, 'Tell us your name', the quick scan through the airwaves lands for half a second on Radio Two, comes up with 'Ken Bruce' and the interviewer, open-mouthed, says to his colleagues, 'Did they just say Ken Bruce?' This is a slight exaggeration, obviously, but one cannot dent the fact that these machines pick up snippets of noise on the airwaves which can then be misinterpreted by the listener.

At the end of the day, none of this is any good if it is not evidence-based.

However, all www.assap.ac.uk can suggest is falling back on the interview, believing that the first-person telling is the closest that you can get to evidence without actually seeing an apparition before your very eyes – and even then it needs two of you to back up each other's stories. The more people that see something, the better the evidence.

The problem is, yes, you've guessed it; people lie.

Saying all this though, probably gives the false impression that I don't believe any of this. Once again, I want to. I think that I have seen things on these programmes that might well be genuine. The problems with this though are so many that I cannot put myself in the position of believer - the editing is often crafty and messes with time-spans and reactions, what happens off-camera is as important as what happens on-camera and we don't see 99% of what happens off-camera. The list could go on and on.

So, the short, tall, fat and thin of this section is that I have nothing to offer you. Ghosts might be real. Ghosts might not be real.

I think that there is so much that we don't know about this body that we inhabit and this earth upon which we live, such as the recent quantum theories that have been put forward (so over my head that they have created a bald patch), that I remain open-minded and hopeful that this, this day to day drudgery, this ant-like existence, is not all there is.

Mediums

Last night I watched a discussion between Derren Brown

and Richard Dawkins on YouTube[304]. Richard Dawkins was surprisingly posh, friendly-looking and short – unless Derren Brown is a titan. I expected him, after *The God Delusion*, to have bits of spittle at the edge of his mouth and the kind eyes that those dogs that you can never quite trust have. You know the ones.

He wasn't like that. He was calm and sensible and fascinating and bore his opinions and scepticism in the way one long-practiced at such things would do. I liked him.

Derren Brown held his anger back, I felt. He was clearly annoyed at the fraud of God and of ghosts and would, I feel, have probably given a world-class, Olympic standard rant given half the chance.

A couple of days prior to this I had watched Brown's new show on Channel Four. He was mind-boggling. Brilliant. During it he did his version of spiritual healing, but he did it with the openly shared knowledge that he was not healing through God or the spirits. He was simply playing psychological tricks. I have to say, I was mesmerised (which is an appropriate word under the circumstances).

Mr Dawkins and Mr Brown were discussing mediums. If you don't know of Derren Brown you will wonder why this should be of any note. Well, our Derren does not believe in mediums. I would go so far as to say that, if he was Prime Minister or King, that we would develop around this country medium holes in the same way that we developed priest holes in the late medieval period. They would hide for their lives.

The thing is, I agreed with them both (and both of them came from educated, logical, experienced angles). Except in one thing. I hold out hope that not all mediums are fraudulent. These two were in no doubt that all mediums were not only fraudulent but were harmful.

> 'If you have lost somebody dear to you and I'm trampling all over those memories by telling you that they are here…saying this and that…that I find

---

304 https://www.youtube.com/watch?v=X56Kmbgn6dE

really quite ugly.[305]'

They both recalled numerous times when they had caught out fake mediums and had seen their blatant 'amateurish' techniques, described by Brown as no better than a cheap stage hypnotist in Corfu.

> 'Many mediums…come from show business stock and have learned their skill, undeniably…I know the teacher of one of the very famous modern psychics who's absolutely taught them his fraudulent techniques and is upset with him for passing them off as real.[306]'

There was one private reading being done with an individual, recounts Brown. The reader said to him that the subject either lived on his own or with a group of other people. I have no psychic training, but even I could have come up with that (with practice). The subject asked the medium to be more specific. What did he mean by this statement? Of course, he lived either on his own or with a group of people. The medium got flustered and the interview was stopped by his advisor, who angrily accused the subject of blocking the medium's psychic channels, an oft used technique apparently.

So what about all these programmes on the TV? Are they committing bare-faced fraud, knowing full well that they are going to drag the money in by the barrel load with advertising, books, T-shirts, pens and mugs? Is there the temptation for the individual, anywhere within the show, from producer, to medium to cameraman, to trump up a bit of evidence to keep the rating and the reputation up? Yep.This article first appeared in The Daily Mirror on 28th Oct 2005

> The TV show that has spooked millions with its

---

[305] Derren Brown. *Richard Dawkins interviews Illusionist Derren Brown (Enemies of Reason Uncut Interviews).* https://www.youtube.com/watch?v=X56Kmbgn6dE

footage of hauntings and poltergeists has been exposed as a fake - by one of its own stars.

Resident parapsychologist Dr Ciaran O'Keeffe has sensationally lifted the lid on the ghosthunting series, Most Haunted ... and claims that the public are being deceived by "showmanship and dramatics."

He accuses the show's medium Derek Acorah of hoodwinking viewers by pretending to communicate with spirits and obtaining information about locations prior to filming.

The Mirror has also obtained unedited footage which appears to show presenter Yvette Fielding and her husband faking 'paranormal' occurrences such as ghostly bumps and knocks.

Yvette Fielding has said of the show - which is made by her husband Karl Beattie's production company Antix Productions: "There is no acting in this programme, none whatsoever. Everything you see and you hear is real. It's not made up, it's not acted."

But our investigation reveals how the programme uses careful editing to mislead viewers and, on at least one occasion, has even lied about the location of filming.

It was disappointing, to say the least. People trusted Acorah and the *Most Haunted* team. The article goes on to describe fraudulent moments and Dr O'Keefe's feelings about the medium. However, there is the advisory at the beginning of the show which says that it is for entertainment purposes only. That is, in fairness, a bit of a warning that something might be a bit skew-whiff.

I still watch *Most Haunted* and probably will continue to do so until I become the invisible star. But I watch it with reserve.

This section is not intended to be an exposé. It is supposed to be an impartial examination into the evidence either for or against the paranormal.

The programmes that I watch continue to produce shadows and photos and EVPs that they say are evidence of something somewhere. I would like to take their word for it, really I would. There is very little I can think of that would please me more than solid evidence of the paranormal and life beyond life (winning the lottery is one).

You can guess where the stronger, or only, evidence lies. In time we might gain more knowledge about ourselves and our environment, until then, someone has to keep plugging on. Why? I don't know.

I still am an avid watcher of Most Haunted. Through each show I will swear and tut and generally act grumpily with impatience, but I will hold out hope of their integrity and eventual success. They did indeed find something, an apparition, at Wentworth Woodhouse, in 2017 and, in fairness, have gone on the Reality TV website to give a more in-depth examination of the apparition. It remains fascinating, but inconclusive.

In the End…

Well, there you go, that's it.

I hope you have been enlightened and frightened and perhaps a little entertained. There has probably been more Death than Disease but, if I'm being entirely honest, and I do try to be most of the time, I was kind of more focused upon the whole death thing. Disease is merely a car for the journey, but a fascinating (I use that word too much, I know) mode of leaving, all the same.

This book could have gone on forever. There was so much more to say about everything in here but, as I keep reminding you, this is only *A Beginner's Guide...* I'd like you to go on to find out more, maybe even find me that proof of the afterlife for which I have been looking or a find a cure for my tinnitus (stood too close to the speakers at a Uriah Heep gig. Worth it).

There are so many ways to shuffle off, like the Canadian lawyer who died while trying to prove that the glass in the windows of a 24th floor office was unbreakable, by throwing himself against it. It didn't break - but it did pop out of its frame and he plunged to his death, the Brazilian who was killed in 2013 when a cow fell through his roof onto him as he slept or the man who became over-excited watching Avatar and had a stroke.

I haven't mentioned Stone Man Syndrome, Tree Man Syndrome, Werewolf Syndrome, Alice in Wonderland Syndrome or Alien Hand Syndrome. There are a whole mass of syndromes out there, some of which haven't yet been discovered. I shall call one of them Undiscovered Syndrome.

Maybe next time.

I'd like to thank all the people who have done all that research out there and continue to do so. I stand upon the shoulders of giants and get a marvellous view.

I'd also like to thank Sally (as ever) for reading through this time and again and putting me right where I have often gone wildly wrong.

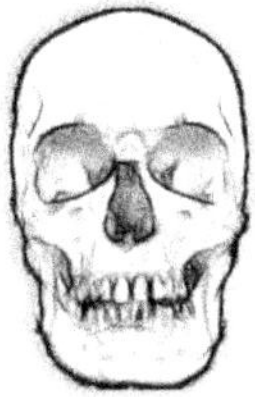

Chris Bradbury
July - October 2016
Reviewed December 2018

ABOUT THE AUTHOR

Chris Bradbury was born in 1962. He attended schools in Bracknell, Windsor, Mauritius and Bloxham.

He has been a shop worker, a hospital porter, worked in medical records, in the CSSD department of a hospital, as an estate agent, as a nurse, as a delivery driver, a bus driver and as a teaching assistant. At the moment he works in a warehouse.

He lives in Barnsley, South Yorkshire, is married and has three children.

He is the author of:

No Time To Repent
The High Commissioner's Wife
The Devil Inside
Catfish
Eidolon
The Stilling of the Heart
Phoenix (a perspective on the causes of WWII)
Shorts – A collection of novellas
Just the Words (a collection of poetry)
Condition of Life - The Poetic Confessions of a Grumpy Old Man
The Ghost of Dormy Place and Other Tales
The Ashes of an Oak
A Kind and Gentle Man
Praxis (Sci-Fi Fantasy - with Ian Makinson)
Praxis – Part Two: Regeneration Paradox (Sci-Fi Fantasy - with Ian Makinson)
Praxis - Part 3 The Liar's truth - (Sci-Fi Fantasy - with Ian Makinson)
Earthbound
Earthbound Part Two - Hellbound
Chine (Horror)
Uncomfortably Numb (Play)
The Scarlet Darter (fiction for children)
Unton's Teeth and Other Tales of Wordful Mystification (poems for children)

A Beginner's Guide to the Wars of the Roses
A Beginner's Guide to Creative Writing

www.ingramcontent.com/pod-product-compliance
Lightning Source LLC
Chambersburg PA
CBHW070312190526
45169CB00005B/1600
* 9 7 8 1 5 3 9 5 2 9 5 2 1 *